Work-units in Arithmetic

Other books in Arithmetic from Stanley Thornes:

A. Greer	*ARITHMETIC FOR COMMERCE* Second Edition
Ewart Smith	*A NEW GENERAL ARITHMETIC*
Ewart Smith	*EXAMPLES IN ARITHMETIC*
Ewart Smith	*EASIER EXAMPLES IN ARITHMETIC*

Work-units in Arithmetic

S. Moseley BSc FEdTC
Peterborough Technical College

Stanley Thornes (Publishers) Ltd

First published in 1986 by:
Stanley Thornes (Publishers) Ltd
Old Station Drive
Leckhampton
CHELTENHAM GL53 0DN
England

British Library Cataloguing in Publication Data

Moseley, S.
Work-units in arithmetic.
1. Arithmetic—Examinations, questions, etc.
I. Title
513'.076 QA139

ISBN 0-85950-513-8

Typeset in 11/13 Garamond by Tech-Set, Gateshead, Tyne & Wear.
Printed and bound in Great Britain at The Bath Press, Avon.

Contents

Preface

These units form a complete course in arithmetic including all the topics needed for students studying for RSA Arithmetic Stage I. Units 1 to 13 contain the material for Paper 1 and units 14 to 20 complete the course.

The work can also be used by students following another first examination course and needing extra help in arithmetic.

The units are written so that students can work through them at their own pace. Many of the examples ask the student to fill in answers within the text. This can be done in the book or on ordinary paper. A short test at the end of each unit enables the student or teacher to monitor progress.

The course was developed for use in the Mathematics Workshop at Peterborough Technical College. I would like to thank the staff and students who used my first version of the text for their encouragement and comments.

The timetables are reproduced by kind permission of the British Railways Board.

My thanks to the RSA for permission to reproduce questions from past RSA Arithmetic Stage I papers. The solutions are my own and not those of the Board.

S. Moseley
Peterborough 1986

To the Student

Read each section thoroughly and follow each example carefully. Then do the examples given. Mark each section immediately you finish it. You will find the answers at the end of each unit. If you have made some mistakes then re-read the section and try again.

When you have completed the whole unit do the test without help. Mark the test, then enter your mark on the progress sheet. If you have made some mistakes try again before going on to the next unit.

Be persistent. If at first you don't succeed, try again.

GOOD LUCK!

Arithmetic Stage I

PROGRESS SHEET

NAME: ______________________

Unit	*Title*	*Date completed*	*Test mark*	*Comments*
1	Number			
2	Decimals			
3	Fractions			
4	Money			
5	Weights and Measures			
6	Magnitude			
7	Percentages			
8	Averages			
9	Ratio and Proportion			
10	Areas			
11	Perimeter and Volume			
12	Time			
13	Revision			
14	Calculators			
15	Profit and Loss			
16	Your Money			
17	Household Expenditure			
18	Boxes, Cans and Prisms			
19	Graphs and Charts			
20	Revision			

UNIT ONE

The Four Rules of Number

A Addition

It is easy to make a careless error when you are doing addition. The following points will help you to avoid these mistakes:

Set your work out neatly in columns.

Start adding on the right.

Put a small number under the line when you are carrying tens to the next column.

Check each sum before going on to the next.

Example

349+ ← *Start here.*
628 *Add from top to bottom.*
977 *9 + 8 = 17*
1 ← *Carry 10 forward to next column.*

You can add more than two numbers at once.

Example

614+
201
17
832

Now you finish these examples:

613+ **284**
———

315+ **228** **197**
———

Check your answers by adding each column from bottom to top.

Now look at the answer (p.8) to see if you were right. If you get any questions wrong correct them if you can. If you do not understand then ask for help.

Try these examples on a sheet of paper:

1. 24 + 17 =	6. 1140 + 869 =
2. 111 + 92 =	7. 14 + 36 + 58 =
3. 19 + 36 =	8. 531 + 137 + 248 =
4. 25 + 114 =	9. 769 + 876 + 968 =
5. 281 + 362=	10. 217 + 5521 + 76 =

B Subtraction

Subtraction means 'take away', e.g. $8 - 3 = 5$. When bigger numbers are involved, there are two methods of subtracting. Both methods give the same answer. Use the method you are used to, or the one you think is easier.
If you are not sure of either method then I suggest you use Method 1.

Example

Method 1

```
4
53-
17
36
```

3–7 we cannot do.
Borrow 10 from 50, cross out 5 and make it 4.
Make 3 into 13.
13 – 7 = 6
4 – 1 = 3

Method 2

```
 1
 53-
117
 36
```

3–7 we cannot do.
Borrow 10 and pay it back.
Make 3 into 13.
Add 1 to the 1 making 2.
13 – 7 = 6
5 – 2 = 3

Check your work by adding the answer to the number you took away.

Check:

```
17+
36
53
```

This gives 53, the number we started with.

Finish these examples, including the checks:

```
28-      191-     1249-
16        72       683
___      ___      ____

16+       72+      683+
___      ___      ____
___      ___      ____
```

Now look at the answers to see if you were right. If you got any questions wrong correct them if you can. If you do not understand, then ask for help.

Try these examples on a sheet of paper. Don't forget to check your answers:

1. 76 – 23 =
2. 95 – 79 =
3. 114 – 96 =
4. 315 – 199 =
5. 444 – 99 =
6. 1984 – 92 =
7. 382 – 94 =
8. 6049 – 581 =
9. 1000 – 981 =
10. 2468 – 1359 =

C Multiplication Tables

×	0	1	2	3	4	5	6	7	8	9	10
0	0	0	0	0	0	0	0	0	0	0	0
1	0	1	2	3	4	5	6	7	8	9	10
2	0	2	4	6	8	10	12	14	16	18	20
3	0	3	6	9	12	15	18	21	24	27	30
4	0	4	8	12	16	20	24	28	32	36	40
5	0	5	10	15	20	25	30	35	40	45	50
6	0	6	12	18	24	30	36	42	48	54	60
7	0	7	14	21	28	35	42	49	56	63	70
8	0	8	16	24	32	40	48	56	64	72	80
9	0	9	18	27	36	45	54	63	72	81	90
10	0	10	20	30	40	50	60	70	80	90	100

Multiplication is repeated addition.
4 × 3 means 3 + 3 + 3 + 3 (4 lots of 3). We can find it on the table by going across from the 4 and down from the 3. We get 12.

Can you see some patterns in the table?

The 0 row is all zeros.
The 5 row has all numbers ending in 0 or 5.
In the 9 row the tens go up 1, 2, 3 . . . while the units go down 9, 8, 7 . . .

Can you find any more patterns?
You may find these patterns helpful when you are trying to remember the tables.

Do you know your tables? Try to fill in the answers from memory:

1. 7 × 3 =
2. 6 × 4 =
3. 8 × 9 =
4. 5 × 8 =
5. 6 × 3 =
6. 2 × 9 =
7. 5 × 8 =
8. 7 × 7 =
9. 9 × 9 =
10. 6 × 7 =

Mark your work. If you have got them all right go on to the next section, division.
If not, copy the table on to a sheet of paper. Stick it on cardboard and keep it in your book. Use it to help you do your work. You will find you learn the tables by using them.

D Division

Division is the reverse of multiplication.

6 × 3 = 18   18 ÷ 6 = 3
(How many 6s make 18?)

Fill in the answers. If you get stuck use your multiplication table:

1. 21 ÷ 7 =   (7 × ? = 21)
2. 70 ÷ 10 =   (10 × ? = 70)
3. 36 ÷ 6 =
4. 32 ÷ 8 =
5. 16 ÷ 2 =
6. 21 ÷ 3 =
7. 45 ÷ 5 =
8. 81 ÷ 9 =
9. 36 ÷ 4 =
10. 72 ÷ 8 =

Now look at the answers to see if you were right. If you have got any questions wrong correct them if you can. If you do not understand then ask for help.

E Short Multiplication

This is the method you use to multiply any number by a whole number from 1 to 9:

Example

```
 175×
   9
1575
 64
```

Method

Set the work out neatly.
9 × 5 = 45

Carry the 4 and write 5 in the answer.
9 × 7 = 63

Add the 4 you have carried.
63 + 4 = 67

Carry the 6 and write 7 in the answer.
9 × 1 = 9
9 + 6 = 15

There are no more numbers to multiply so write 15 in the answer.

Check your working.

Now try these examples:

```
23×
  7
___

___
```

```
132×
   3
____

____
```

```
876×
   8
____

____
```

Look at the answers as before and correct if necessary. Then try these examples on paper:

1. 22 × 6 =	6. 29 × 3 =
2. 28 × 7 =	7. 54 × 8 =
3. 94 × 5 =	8. 39 × 7 =
4. 53 × 9 =	9. 211 × 6 =
5. 63 × 4 =	10. 581 × 9 =

F Multiplication by Tens

3 × 10 = 30 To multiply by 10 we add a 0
3 × 100 = 300 To multiply by 100 we add 00
3 × 1000 = 3000 To multiply by 1000 we add 000

Fill in the answers:

1. 4 × 10 =	6. 10 × 20 =
2. 5 × 100 =	7. 1000 × 37 =
3. 7 × 10 =	8. 100 × 12 =
4. 32 × 10 =	9. 100 × 10 =
5. 15 × 100 =	10. 100 × 100 =

Mark and correct your work.

G Long Multiplication

First we will multiply by numbers ending in zeros.

Example

```
 15×
 20
----
300
----
 1
```

20 = 2 × 10
(We multiply by 2 and add a 0).
Put down the 0 first then multiply by 2.

Example

```
171×
400
----
 00
----
```

Write 00.
Multiply by 4.
Can you finish it?

If you got 68 400 then do these on paper:

1. 18 × 30 =	4. 419 × 500 =
2. 28 × 200 =	5. 28 × 2000 =
3. 16 × 80 =	

When you have done these correctly study the next example carefully:

Example

```
 213×
  23
----
 639
4260
----
4899
----
```

Set the work out neatly in columns.
Multiply by 3.
Multiply by 20, put a 0 and multiply by 2. Add.

Fill in these examples:

```
 52×
 21
----

----
----
```

```
114×
 45
----

----
----
```

Look at the answers to see if you were right. Ask for help if you do not understand the method. Then do these examples neatly on paper:

1. 25 × 18 =	6. 197 × 56 =
2. 71 × 29 =	7. 243 × 29 =
3. 62 × 43 =	8. 362 × 43 =
4. 827 × 94 =	9. 481 × 58 =
5. 718 × 91 =	10. 845 × 56 =

Did you remember to check each one as you did it? Mark your work and correct any you have got wrong.

H Short Division

This is the method to use if the number you are dividing by is less than 10.

Example

```
   72
8)5¹76
```

Note the position of the numbers.
8 goes into 57 seven times remainder 1.
Write 7 and carry 1.
8 goes into 16 two times with no remainder.
Write 2.

Fill in these examples:

7)392   **5)2165**

When you have marked them try these on paper:

1. $216 \div 8 =$	6. $336 \div 7 =$
2. $356 \div 4 =$	7. $365 \div 5 =$
3. $204 \div 6 =$	8. $111 \div 3 =$
4. $84 \div 6 =$	9. $624 \div 8 =$
5. $243 \div 9 =$	10. $462 \div 3 =$

Now look at the answers to see if you were right. If you got any questions wrong correct them if you can. If you do not understand then ask for help.

I Division by Tens

$300 \div 10 = 30$ To divide by 10 cross off a 0
$300 \div 100 = 3$ To divide by 100 cross off 00
Fill in the answers:

1. $50 \div 10 =$	4. $5000 \div 100 =$
2. $200 \div 10 =$	5. $550 \div 10 =$
3. $200 \div 100 =$	

J Long Division

It is very important to keep your work in neat columns.

Example

```
     25
 23)575
    46
    ---
    115
    115
    ---
      0
```

23 will not go into 5.
23 goes into 57 twice, write 2.
$23 \times 2 = 46$
$57 - 46 = 11$
There are 11 left over.
Bring down the 5.
23 goes into 115 five times, write 5.
$23 \times 5 = 115$
$115 - 115 = 0$
There is no remainder.

Rough work

```
 23×      23×
  2        3
 --       --
 46       69

 23×      23×
  4        5
 --      ---
 92      115
 1        1
```

Check your multiplication by 23 in a column labelled rough work. If necessary, multiply 23 by every number from 1 to 9 before you start to divide.

We shall see what to do with remainders when we get to decimals. You finish these examples:

63)4410

Rough work

```
 63×
  7
 ---
 441
  2
```

```
      1
25)275          18)3024

    5                 2

    0                 4

                      0
```

When you have marked and if necessary corrected them, try these:

1. 78 ÷ 13 =	6. 609 ÷ 29 =
2. 108 ÷ 27 =	7. 7125 ÷ 95 =
3. 800 ÷ 32 =	8. 8823 ÷ 51 =
4. 437 ÷ 23 =	9. 9270 ÷ 45 =
5. 7308 ÷ 36 =	10. 12 879 ÷ 81 =

K Series

If you are asked to complete a series of numbers look for a pattern.

Example

Complete the series: 2, 4, 6, 8, __, __
You may recognise this series as coming from the 2 times table.
The numbers increase by 2 each time.

2 4 6 8 10 12
+2 +2 +2 +2 +2

So the next two numbers are 10, 12.

Example

Complete the series: 9, 10, 12, 15, 19, __, __
This series is not so easily recognisable.
But look at this:

9 10 12 15 19
+1 +2 +3 +4

Now you can see the pattern.
The next two numbers are

19 + 5 = **24**
and 24 + 6 = **30**

Now you try these, completing the series:

1. 1, 2, 3, 4, __, __
2. 3, 5, 7, 9, __, __
3. 10, 9, 8, 7, __, __
4. 19, 21, 23, 25, __, __
5. 80, 85, 90, 95, __, __
6. 4, 6, 9, 13, __, __
7. 16, 19, 22, 25, __, __
8. 9, 19, 28, 36, __, __
9. 6, 8, 12, 18, __, __
10. 2, 4, 8, 16, __, __

Check your work and then mark it from the answers.

This is the end of Unit One. When you are sure you have understood it try the test exercise without looking at your notes. Hand it in to be marked and go on to Unit Two.

UNIT ONE **Test**

Set your work out neatly on lined paper. Put your name at the top.

1. 628 + 4190 + 36 =
2. 491 − 216 =
3. 49 × 7 =
4. 73 × 10 =
5. 26 × 100 =
6. Complete the series 7, 11, 15, 19, __, __
7. 3600 ÷ 100 =
8. 68 × 24 =
9. 2665 ÷ 65 =
10. 630 ÷ 9 =

Do not forget to check your work.

A

```
 613+        315+
 284         228
 ---         197
 897         ---
             740
             1 2
```

1.	41	2.	203	3.	55
4.	139	5.	643	6.	2009
7.	108	8.	916	9.	2613
10.	5814				

B

```
              8 1                 1
 28−          191−     or       191−
 16            72               ¹72
 --           ---               ---
 12           119               119

 16+                  72+
 12                  119
 --                  ---
 28                  191
                       1

  1 1              1 1
 ⁰1249−    or     1249
   683            ₁¹683
  ----            ----
   566             566

        683+
        566
        ----
        1249
         1 1
```

1.	53	2.	16	3.	18
4.	116	5.	345	6.	1892
7.	288	8.	5468	9.	19
10.	1109				

C

1.	21	2.	24	3.	72
4.	40	5.	18	6.	18
7.	40	8.	49	9.	81
10.	42				

D

1.	3	2.	7	3.	6
4.	4	5.	8	6.	7
7.	9	8.	9	9.	9
10.	9				

E

```
  23×        132×        876×
   7           3            8
 ---         ---         ----
 161         396         7008
  2                       6 4
```

1.	132	2.	196	3.	470
4.	477	5.	252	6.	87
7.	432	8.	273	9.	1266
10.	5229				

F

1.	40	2.	500	3.	70
4.	320	5.	1500	6.	200
7.	37 000	8.	1200	9.	1000
10.	10 000				

G

1.	540	2.	5600	3.	1280
4.	209 500	5.	56 000		

```
   52×          114×
   21             45
 ----           ----
   52            570
 1040           4560
 ----           ----
 1092           5130
```

1.	450	2.	2059	3.	2666
4.	77 738	5.	65 338	6.	11 032
7.	7047	8.	15 566	9.	27 898
10.	47 320				

H

```
   56
 7)392   (carry 4)
```

```
   433
 5)2165   (carries 1 1)
```

1. 27 2. 89 3. 34
4. 14 5. 27 6. 48
7. 73 8. 37 9. 78
10. 154

I

1. 5 2. 20 3. 2
4. 50 5. 55

J

```
     70
 63)4410
    441
     00
```

```
     11
 25)275
    25
     25
     25
      0
```

```
     168
 18)3024
    18
    122
    108
     144
     144
       0
```

Rough work

```
 18×
   6
 108
```

```
 18×
   8
 144
```

1. 6 2. 4 3. 25
4. 19 5. 203 6. 21
7. 75 8. 173 9. 206
10. 159

K

1. 5, 6 2. 11, 13 3. 6, 5
4. 27, 29 5. 100, 105 6. 18, 24
7. 28, 31 8. 43, 49 9. 26, 36
10. 32, 64

UNIT TWO

Decimals

The number 365 means 3 hundreds and 6 tens and 5 units.
The number system can be extended into tenths, hundredths etc.
The decimal point separates the whole numbers from the decimal fractions.

H	T	U	t	h
3	6	5 .	2	4

365.24 means 3 hundreds and 6 tens and 5 units and 2 tenths and 4 hundredths.

A Addition and Subtraction

The rule is:
Keep the points underneath each other. Fill in the gaps at either end of the number with zeros. Follow the method for addition and subtraction that you learnt in Unit One.

Example

35.4 + 5.25

Set out the sum in columns with the points under each other. You can add a zero at the end of a decimal number without altering its value. Put a zero before 5.25 to show that there is nothing in the ten's column.

35.40+
05.25
40.65

Example

3 1
35.40−
05.25
30.15

or

1
35.40−
05.[1]25
30.15

Subtract using the method you chose in Unit One. Fill in the answers:

17.61+
4.92
_____

18.58−
4.92
_____

3.09−
0.67
_____

Check your answers and then mark them from the answer page (p. 14). Correct them if necessary and then try these examples. Set them out neatly on paper:

1. 7.1 + 8.2 =
2. 16.3 + 0.81 =
3. 12.6 − 5.4 =
4. 18.5 − 7.62 =
5. 50.13 + 6.083 + 14.95 =
6. 38.03 + 17 + 0.012 =
   (remember 17 is the same as 17.0)
7. 28.813 − 17.41 =
8. 215.06 − 109.59 =
9. 164.92 + 21.795 =
10. 33.33 − 7.775 =

Mark and correct your work as usual.

B Multiplication by Tens

2 × 10 = 20

can be written

2.0 × 10 = 20.0

The decimal point appears to have moved one place to the right.

2.0 × 100 = 200.0

The decimal point appears to have moved two places to the right. We can use this method to multiply quickly by 10, 100, 1000 etc. Add some zeros to the number if necessary.

Examples

1.86 × 10 = 18.6

1.86 × 100 = 186.0 = 186

3.2 × 100 = 320

You fill in the answers:

1. 2.4 × 10 =	6. 95.25 × 100 =
2. 3.54 × 10 =	7. 26.521 × 1000
3. 92.75 × 100 =	8. 26.38 × 1000 =
4. 3.6 × 100 =	9. 0.6 × 100 =
5. 21 × 10 =	10. 3.2 × 1000 =

When you have marked and corrected your work go on to the next section.

C Short Multiplication and Long Multiplication

Decimals are multiplied in the same way as whole numbers, but you must put the decimal point in the correct position in the answer.
Sometimes it is easy to estimate the size of the answer and then position the decimal point.

Example

3.8 × 6

$$\begin{array}{r} 38\times \\ 6 \\ \hline 228 \\ \hline \end{array}$$

3.8 × 6 is nearly 4 × 6 = 24.
The answer is **22.8**.

Another method used to position the decimal point is shown in this example.

Example

19.2 × 0.51

$$\begin{array}{r} 192\times \\ 051 \\ \hline 192 \\ 9600 \\ \hline 9792 \\ \hline \end{array}$$

Multiply the numbers ignoring the point.

Now look at this:

1 + 2 equals 3

19.2 × 0.51 = 9.792

By counting how many figures are after the decimal point, we can put the decimal point in the correct place in the answer.
The answer is 9.792

Now try to do these questions:

$$\begin{array}{r} 9.2\times \\ 0.5 \\ \hline \end{array}$$

$$\begin{array}{r} 17.2\times \\ 3.4 \\ \hline \end{array}$$

$$\begin{array}{r} 14.2\times \\ 3.1 \\ \hline \end{array}$$

Make sure you have put the decimal points in the correct position.

Mark your work and then try these questions on paper:

1. 2.3 × 7 =	6. 2.7 × 1.6 =
2. 9.6 × 0.3 =	7. 3.4 × 1.3 =
3. 8.1 × 4 =	8. 1.201 × 2.3 =
4. 9.7 × 0.02 =	9. 0.32 × 1.44 =
5. 0.2 × 0.2 =	10. 0.06 × 0.04 =

Check and mark your work as before.

D Division by Tens

20 ÷ 10 = 2

can be written

20.0 ÷ 10 = 2.00

The decimal point appears to have moved one place to the left.

20.0 ÷ 100 = 0.200

The decimal point appears to have moved two places to the left.
We can use this method to divide quickly by 10, 100, 1000 etc.
Add some zeros to the number if necessary.

Examples

26.3 ÷ 10 = 2.63

26.3 ÷ 100 = 0.263

3.6 ÷ 100 = 0.036

Fill in the answers.
Look at the signs carefully:

1. 36.9 ÷ 10 =	6. 17.2 × 10 =
2. 21.8 ÷ 100 =	7. 19.6 ÷ 100 =
3. 17 ÷ 10 =	8. 29.7 ÷ 1000 =
4. 82 ÷ 100 =	9. 2.66 × 100 =
5. 9.75 ÷ 100 =	10. 0.4 ÷ 10 =

Mark and correct your work.

E Short Division and Long Division

Move the decimal point of the number you are dividing by to the extreme right.
Move the other point the same number of places.
Divide by short or long division keeping the points in line.
Divide until there is no remainder.

Example

1.68 ÷ 0.7
= 16.8 ÷ 7

```
   2.4
7)16.²8
```

Example

6 ÷ 2.4
= 60 ÷ 24

```
     2.5
24)60.0
   48
   120
   120
     0
```

Fill in these examples:

2.65 ÷ 0.2
= 26.5 ÷ 2
= ANS

2)26.5

7.65 ÷ 1.7
= ÷ 17
= ANS

17)

Check your work and then mark it. If you have made any mistakes write the examples out again. Then try these on paper. Work out each answer exactly or if the answer does not work out exactly, stop when you have three figures after the point:

1. 2.4 ÷ 0.2 =
2. 7.4 ÷ 2 =
3. 65.8 ÷ 0.4 =
4. 7.62 ÷ 0.9 =
5. 9.71 ÷ 0.04 =
6. 1.92 ÷ 1.2 =
7. 2.24 ÷ 1.4 =
8. 3.36 ÷ 2.4 =
9. 0.364 ÷ 0.14 =
10. 350.56 ÷ 5.6 =

Now mark your work and correct any you have got wrong.

F Hidden Decimals

Example

11 ÷ 4

$$4\overline{)11.00}\quad = 2.75$$

(carries: 3, 2)

There are no decimal points in the question but we need them in the answer.

Try these on paper:
Work out each answer exactly or continue until you have two figures after the point.

1. 15 ÷ 4 =
2. 7 ÷ 6 =
3. 220 ÷ 3 =
4. 19 ÷ 7 =
5. 22 ÷ 5 =

Check your work and mark it.

G Mixed Examples

You should now understand decimals.
Try these mixed examples:

1. 3.8 + 7.91 =
2. 78.4 – 16.9 =
3. 8.9 × 100 =
4. 2.75 × 6 =
5. 5.1 × 9.7 =
6. 7.9 ÷ 10 =
7. 16 ÷ 100 =
8. 5.2 ÷ 6 =
9. 1.2 ÷ 0.9 =
10. 6 ÷ 5 =

Mark your work. If there are any questions you are not sure of turn back to the appropriate section or ask for further help.
When you are happy with decimals do the test below.

UNIT TWO **Test**

Set your work out neatly on lined paper. Put your name at the top.

1. 21.4 + 1.79 =
2. 13.6 – 3.21 =
3. 2.9 × 7 =
4. 5.8 × 4.2 =
5. 7.9 × 100 =
6. 8.54 ÷ 100 =
7. 27.3 ÷ 6 =
8. 29.34 ÷ 1.8 =
9. 21 ÷ 4 =
10. 0.6 ÷ 0.15 =

Do not forget to check your work.

A

$$\begin{array}{r} 17.61+ \\ 4.92 \\ \hline 22.53 \end{array} \qquad \begin{array}{r} 18.58- \\ 4.92 \\ \hline 13.66 \end{array} \qquad \begin{array}{r} 3.09- \\ 0.67 \\ \hline 2.42 \end{array}$$

1. 15.3 2. 17.11 3. 7.2
4. 10.88 5. 71.163 6. 55.042
7. 11.403 8. 105.47 9. 186.715
10. 25.555

B

1. 24 2. 35.4 3. 9275
4. 360 5. 210 6. 9525
7. 26 521 8. 26 380 9. 60
10. 3200

C

$$\begin{array}{r} 9.2\times \\ 0.5 \\ \hline 4.60 \end{array} \qquad \begin{array}{r} 17.2\times \\ 3.4 \\ \hline 688 \\ 5160 \\ \hline 58.48 \end{array} \qquad \begin{array}{r} 142\times \\ 3.1 \\ \hline 142 \\ 4260 \\ \hline 440.2 \end{array}$$

1. 16.1 2. 2.88 3. 32.4
4. 0.194 5. 0.04 6. 4.32
7. 4.42 8. 2.7623 9. 0.4608
10. 0.0024

D

1. 3.69 2. 0.218 3. 1.7
4. 0.82 5. 0.0975 6. 172
7. 0.196 8. 0.0297 9. 266
10. 0.04

E

$$\begin{array}{r} 13.25 \\ 2\overline{)26.50} \end{array} \qquad \begin{array}{r} 4.5 \\ 17\overline{)76.5} \\ 68 \\ 85 \\ 85 \\ \hline 00 \end{array}$$

1. 12 2. 3.7 3. 164.5
4. 8.466 5. 242.75 6. 1.6
7. 1.6 8. 1.4 9. 2.6
10. 62.6

F

1. 3.75 2. 1.16 3. 73.33
4. 2.71 5. 4.4

G

1. 11.71 2. 61.5 3. 890
4. 16.5 or 16.50 5. 49.47
6. 0.79 7. 0.16 8. 0.86
9. 1.33 10. 1.2

UNIT THREE

Fractions

As you have now successfully completed two sections of this course the following instructions will not be repeated for every exercise.

Remember to:
Read each section carefully.
Try the 'fill in' examples and then mark them yourself.
If you made any mistakes find out where you went wrong.
If spaces have not been left on the page for your working, then set the exercise out neatly on paper. *Don't try to squash it on the page. Untidy working leads to errors.*
Complete the exercise then mark it from the answer sheet.
Do any corrections that are necessary.
Now do the work on fractions.

A What is a Fraction?

A fraction is part of a *whole one*. The fraction consists of two parts:
the *top* or *numerator* tells us how many parts we have;
the *bottom* or *denominator* tells us the name of the fraction part we are using.

$\frac{1}{2}$ means one half (1 of 2 parts)

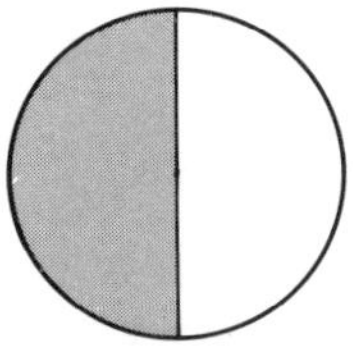

$\frac{3}{4}$ means three quarters (3 of 4 parts)

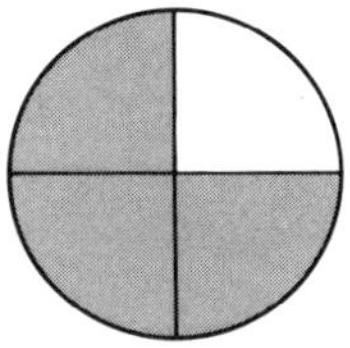

$\frac{5}{8}$ means five eighths (5 of 8 parts)

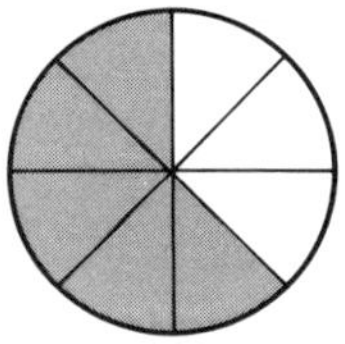

Write next to each diagram the name of the fraction:

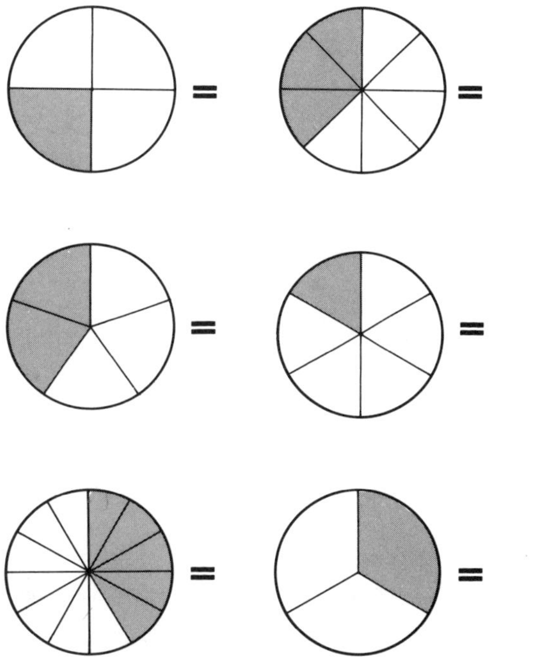

Complete these diagrams to show the given fraction:

$\frac{7}{8}$ = 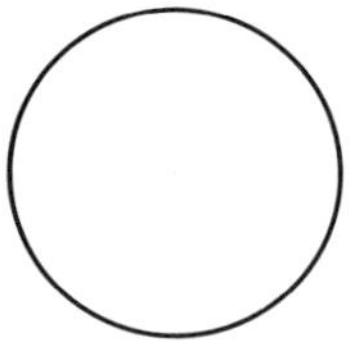 $\frac{3}{5}$ = 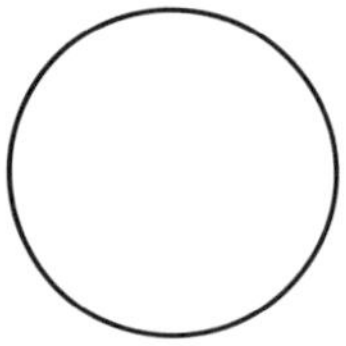

$\frac{4}{5}$ = 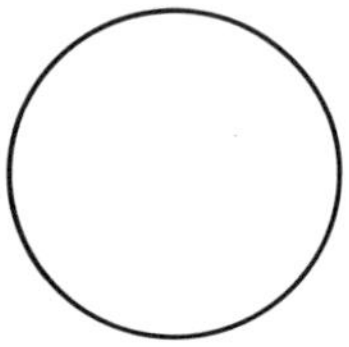 $\frac{7}{12}$ = 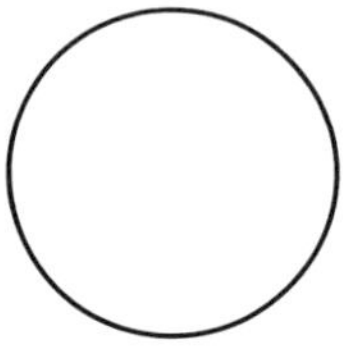

$\frac{2}{3}$ = 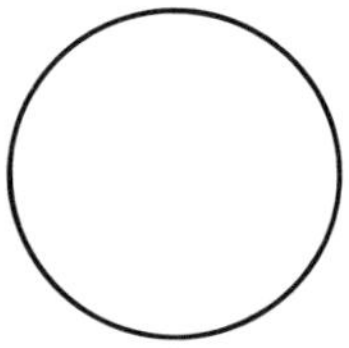 $\frac{2}{9}$ = 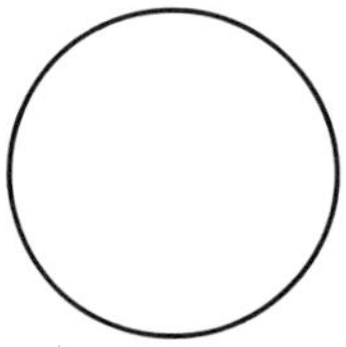

A fraction can also be thought of as a division.

$\frac{2}{3}$ means the result of **2 ÷ 3.**

If you share two cakes between Ann, Belinda and Carol, they get $\frac{2}{3}$ each.

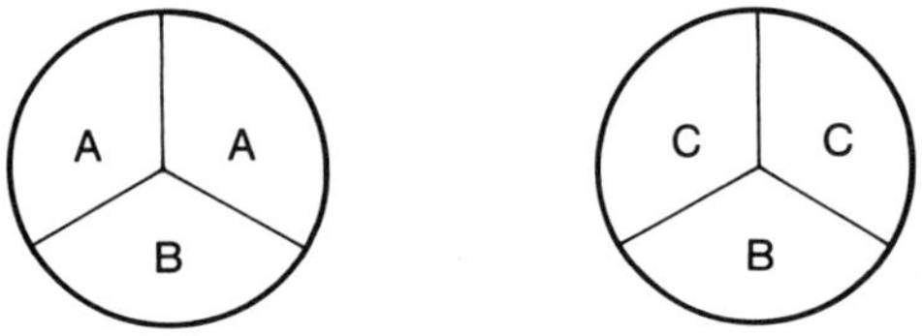

Now let us do $\frac{5}{8}$. This means **5 ÷ 8.**

We will share five cakes between eight boys, Andy, Barry, Charles, David, Edward, Fred, George and Harry. This is what they get: $\frac{5}{8}$ each.

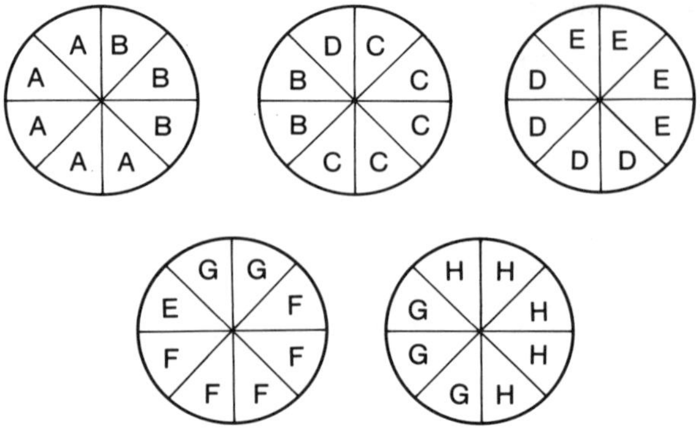

Now you divide these cakes to show $\frac{3}{5}$ **= 3 ÷ 5.**

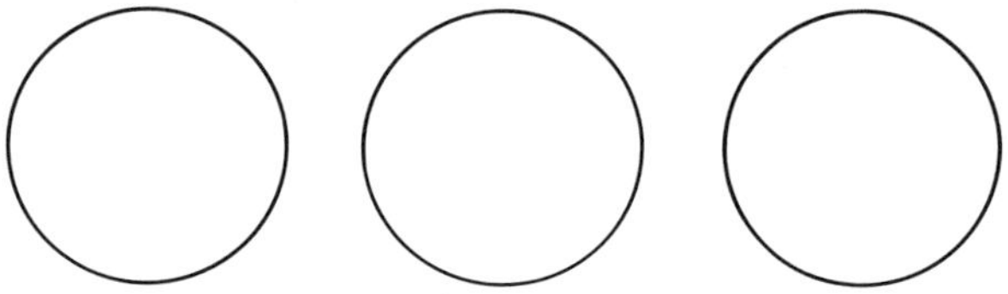

B Equal Fractions

$\frac{1}{2}$ = 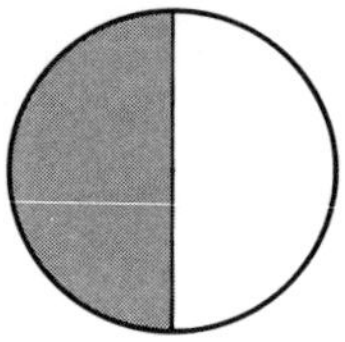 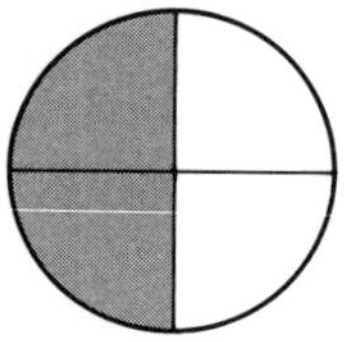 = $\frac{2}{4}$

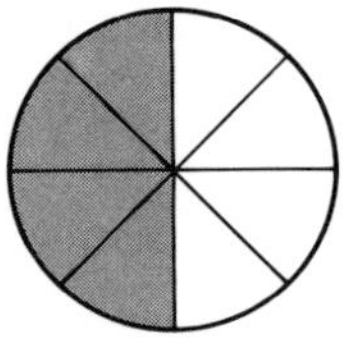 = $\frac{4}{8}$

You can see that $\frac{1}{2}$ is exactly the same as $\frac{2}{4}$ and $\frac{4}{8}$.

When you are working with fractions you can multiply or divide the top and bottom by the same number.

For instance $\frac{2}{4} = \frac{1}{2}$ *Divide the top and bottom by 2.*

$\frac{2}{4} = \frac{4}{8}$ *Multiply the top and bottom by 2.*

You will see later why this is useful.

Example 1

Simplify $\frac{12}{16}$ *This means make the numbers as small as possible.*

$\frac{12}{16} = \frac{3}{4}$ *Divide the top and bottom by 4, or divide the top and bottom by 2 and 2 again.*

$$\frac{12}{16} = \frac{6}{8} = \frac{3}{4}$$

Now you simplify these:

1. $\frac{16}{20} =$
2. $\frac{5}{30} =$
3. $\frac{3}{9} =$
4. $\frac{15}{20} =$
5. $\frac{8}{12} =$
6. $\frac{3}{18} =$
7. $\frac{25}{75} =$
8. $\frac{72}{90} =$
9. $\frac{72}{84} =$
10. $\frac{20}{32} =$

Example 2

Sometimes we need to make the numbers larger so we multiply both top and bottom by the same amount.

Change $\frac{3}{4}$ *to twelfths:*

$$\frac{3}{4} = \frac{3 \times 3}{4 \times 3} = \frac{9}{12}$$

Multiply top and bottom by 3 in this case because 4 × 3 gives 12.

Now you change $\frac{2}{3}$ *so that the bottom is 15:*

$$\frac{2}{3} = \frac{2 \times }{3 \times } = \frac{}{15}$$

Now complete these:

1. $\frac{3}{4} = \frac{}{8}$
2. $\frac{1}{2} = \frac{}{12}$
3. $\frac{3}{4} = \frac{}{20}$
4. $\frac{1}{3} = \frac{}{18}$
5. $\frac{5}{8} = \frac{}{24}$
6. $\frac{11}{20} = \frac{}{80}$

Example 3

Sometimes we need to change a whole number and a fraction (known as a 'mixed number') all into a fraction. This is called an 'improper fraction'. It means a top-heavy fraction.

Change $2\frac{3}{4}$ *to an improper fraction:*

We must change the 2 whole ones to quarters and then add on the $\frac{3}{4}$.

$$2 = \frac{2}{1} = \frac{8}{4}$$

so $2\frac{3}{4} = 2 + \frac{3}{4}$

$$= \frac{8}{4} + \frac{3}{4} = \frac{11}{4}$$

Change $1\frac{2}{3}$ *to an improper fraction:*

We must change the 1 whole one to thirds and then add on the $\frac{2}{3}$.

$$1 = \frac{1}{1} = \frac{3}{3}$$

so $1\frac{2}{3} = \frac{3}{3} + \frac{2}{3} = \frac{5}{3}$

Now you change these to improper fractions:

1. $2\frac{1}{2} =$
2. $6\frac{1}{4} =$
3. $5\frac{1}{9} =$
4. $1\frac{5}{6} =$
5. $2\frac{2}{3} =$
6. $11\frac{1}{3} =$
7. $12\frac{1}{9} =$
8. $5\frac{3}{4} =$
9. $3\frac{3}{8} =$
10. $4\frac{2}{7} =$

Example 4

Sometimes we need to change an improper fraction to a mixed number.

Change $\frac{25}{4}$ *to a mixed number:*

25 ÷ 4 = 6 remainder 1

25 quarters makes 6 whole ones with one quarter left over.

$$\frac{25}{4} = 6\frac{1}{4}$$

$6\frac{1}{4}$ *is a mixed number because it is a whole number and a fraction.*

Change $\frac{17}{3}$ *to a mixed number:*

17 ÷ 3 = 5 remainder 2

17 thirds makes 5 whole ones with two thirds left over.

$$\frac{17}{3} = 5\frac{2}{3}$$

Now you try these:

1. $\frac{20}{9} =$
2. $\frac{17}{4} =$
3. $\frac{20}{5} =$
4. $\frac{19}{3} =$
5. $\frac{5}{2} =$
6. $\frac{15}{4} =$
7. $\frac{21}{5} =$
8. $\frac{32}{7} =$
9. $\frac{22}{11}$
10. $\frac{81}{6} =$

C Addition

Example

$2\frac{1}{2} + 3\frac{3}{4}$ — *2 + 3 = 5. 2 and 4 both divide into 4.*

$= 5\frac{2}{4} + \frac{3}{4}$ — *Change both fractions to quarters. Add.*

$= 5\frac{5}{4} = 6\frac{1}{4}$ — *It is top heavy, so simplify.*

You try these:

$1\frac{1}{2} + 1\frac{1}{4}$

$= \quad \frac{\ }{4} + \frac{\ }{4}$

$=$ ANS

$5\frac{3}{4} + 3\frac{2}{3}$

$= \quad — + —$

$=$ ANS

Try these examples on paper:

1. $1\frac{1}{2} + 1\frac{1}{3} =$
2. $2\frac{1}{5} + 1\frac{1}{10} =$
3. $1\frac{2}{3} + 2\frac{5}{9} =$
4. $4\frac{1}{5} + 1\frac{1}{4} =$
5. $3\frac{2}{3} + 2\frac{3}{4} =$
6. $1\frac{1}{4} + \frac{7}{15} =$
7. $1\frac{3}{16} + 1\frac{5}{8} =$
8. $18\frac{7}{8} + 21\frac{3}{4} =$
9. $114\frac{1}{2} + 215\frac{2}{3} =$
10. $92\frac{2}{5} + 106\frac{2}{7} =$

D Subtraction

Example

$3\frac{2}{3} - 2\frac{1}{4}$ — *3 − 2 = 1. 3 and 4 both divide into 12.*

$= 1\frac{8-3}{12}$ — *Change both fractions to twelfths. Subtract.*

$= 1\frac{5}{12}$ — *It will not simplify.*

Example

$6\frac{1}{3} - 3\frac{1}{2}$ — *6 − 3 = 3. 3 and 2 both divide into 6.*

$= 3\frac{2-3}{6}$ — *Change both fractions to sixths. We cannot take 3 from 2.*

$= 2\frac{6+2-3}{6}$ — *Change one whole one to sixths. 6 − 2 − 3 = 5*

$= 2\frac{5}{6}$ — *It will not simplify.*

Fill in these examples:

$5\frac{1}{2} - 1\frac{1}{3}$

$= \frac{\quad}{6}$

$=$

$4\frac{1}{4} - 2\frac{2}{3}$

$= \frac{\quad}{\quad}$

$= \frac{\quad}{\quad}$

Try these examples on paper:

1. $2\frac{1}{2} - 1\frac{1}{3} =$	6. $8\frac{1}{2} - 6\frac{2}{3} =$
2. $3\frac{3}{4} - 1\frac{5}{8} =$	7. $13\frac{1}{6} - 2\frac{4}{9} =$
3. $1\frac{7}{20} - \frac{14}{15} =$	8. $3\frac{2}{3} - 1\frac{3}{8} =$
4. $5\frac{2}{3} - 3\frac{1}{3} =$	9. $72\frac{1}{2} - 15\frac{3}{8} =$
5. $5\frac{1}{3} - 2\frac{1}{2} =$	10. $131\frac{2}{9} - 72\frac{3}{4} =$

E Multiplication

Example

$\frac{2}{3} \times 3\frac{1}{2}$ — *Change to improper fractions.*

$= \frac{\not{2}^{1}}{3} \times \frac{7}{\not{2}_{1}}$ — *Cancel: this means divide one number on top and one on the bottom by the same thing. We have divided by 2.*

$= \frac{7}{3} = 2\frac{1}{3}$ — *Multiply the top then the bottom. Simplify.*

Example

$\frac{7}{15} \times 1\frac{3}{7}$ — *Change to improper fractions.*

$= \frac{{}^{1}\not{7}}{{}_{3}\not{15}} \times \frac{\not{10}^{2}}{\not{7}_{1}}$ — *Cancel a) 7 and 7 by 7*
b) 10 and 15 by 5.

$= \frac{2}{3}$ — *Multiply the top then the bottom. It will not simplify further.*

Remember 'of' means multiply so $\frac{3}{4}$ of 12 means $\frac{3}{4} \times 12$.

$= \frac{3}{{}_{1}\not{4}} \times \frac{\not{12}^{3}}{1} = \frac{9}{1} = 9$

Now you fill in the answers to these:

$1\frac{1}{2} \times 1\frac{1}{3}$

$= \frac{\quad}{2} \times \frac{\quad}{3}$

$=$ ANS

$1\frac{3}{10} \times \frac{5}{8}$

$= \frac{\quad}{\quad} \times \frac{\quad}{\quad}$

$=$ ANS

Try these examples on paper:

1. $1\frac{1}{2} \times 3 =$	6. $4\frac{1}{6} \times 1\frac{1}{15} =$
2. $1\frac{1}{2} \times 2\frac{1}{4} =$	7. $5\frac{1}{2} \times \frac{3}{8} =$
3. $3\frac{1}{3} \times \frac{3}{5} =$	8. $3\frac{1}{4} \times 1\frac{1}{3} =$
4. $\frac{2}{3} \times \frac{3}{5} =$	9. $4\frac{1}{4} \times 1\frac{1}{2} =$
5. $\frac{9}{11} \times 1\frac{1}{10} =$	10. $\frac{5}{11} \times 4\frac{2}{5} =$

F Division

Change the $\div$ to $\times$ and turn the *second* fraction upside down.

Example

$4\frac{1}{6} \div 3\frac{3}{4}$ — *Change to improper fractions.*

$= \frac{25}{6} \div \frac{15}{4}$ — *Change the sign to × and turn the second fraction upside down.*

$= \frac{{}^{5}\cancel{25}}{{}_{3}\cancel{6}} \times \frac{\cancel{4}^{2}}{\cancel{15}_{3}}$ — *Cancel a) 25 and 15 by 5*
b) 6 and 4 by 2.

$= \frac{10}{9} = 1\frac{1}{9}$ — *Multiply the top then the bottom.*
Simplify.

Fill in these examples:

$1\frac{1}{2} \div 2\frac{1}{4}$	$3\frac{1}{7} \div 2\frac{5}{14}$
$= \frac{}{2} \div \frac{}{4}$	$= \frac{}{7} \div \frac{}{14}$
$= - \times -$	$= - \times -$
$=$ ANS	$=$ ANS

Try these yourself on paper:

1. $1\frac{1}{2} \div 3 =$	6. $2\frac{3}{5} \div \frac{4}{5} =$
2. $3\frac{1}{3} \div 1\frac{2}{3} =$	7. $2\frac{2}{9} \div 1\frac{2}{3} =$
3. $3\frac{1}{5} \div 1\frac{7}{25} =$	8. $\frac{4}{5} \div 4 =$
4. $2\frac{1}{4} \div 3 =$	9. $8 \div \frac{2}{3} =$
5. $5\frac{1}{2} \div \frac{4}{5} =$	10. $12\frac{2}{9} \div 1\frac{9}{11} =$

G Decimals to Fractions

Example

Change 0.84 to a fraction and simplify:

$\mathbf{0.84 = \frac{84}{100}}$ — *There are two figures after the point so it is 84 hundredths.*
There are two zeros on the bottom.

$\mathbf{= \frac{21}{25}}$ — *Cancel by 4.*

$\mathbf{0.6 = \frac{6}{10}}$ — *Here there is one figure after the point so one zero on the bottom.*

$\mathbf{= \frac{3}{5}}$ — *Cancel by 2.*

$$\mathbf{20.125 = 20\frac{125}{1000} = 20\frac{1}{8}}$$

The 20 is a whole number so it will not change.

Complete these:

$0.65 = \frac{}{100} =$ $\qquad$ $2.175 = 2 \frac{}{} =$

Now change these to fractions and cancel as much as possible:

1.	0.5	6.	3.675
2.	0.25	7.	9.15
3.	3.125	8.	15.36
4.	5.8	9.	121.75
5.	17.2	10.	100.7

H Fractions to Decimals

You have already done these in Unit Two. They were called Hidden Decimals.

Just to remind you:

$$\begin{array}{r} 0.625 \\ 8\overline{)5.\overset{5}{0}\overset{2}{0}\overset{4}{0}} \end{array}$$

$\frac{5}{8} = 5 \div 8 = 0.625$

Complete this one $\frac{4}{5} =$ ANS

If you get it wrong try the questions in Unit Two F again.

I Now See if You Can Do Mixed-up Fractions

If you are not sure how to do a question use your previous work to help you. You will learn the methods by referring back when necessary, don't guess:

1. $3\frac{1}{2} + 2\frac{1}{5} =$
2. $4\frac{1}{4} - 3\frac{1}{3} =$
3. $\frac{1}{4} \times \frac{2}{3} =$
4. Change $\frac{3}{4}$ to a decimal
5. $7\frac{1}{2} \div 3$
6. Change 0.65 to a fraction
7. Change $\frac{3}{4}$ to twelfths
8. Simplify $\frac{25}{100}$
9. Find $\frac{2}{3}$ of 18
10. $\frac{2}{10} =$

Check the answers before you continue.

11. $\frac{5}{11} + \frac{2}{5} =$
12. $3\frac{1}{3} - 2\frac{1}{2} =$
13. $1\frac{1}{3} \times 1\frac{1}{4} =$
14. Change $\frac{3}{8}$ to a decimal
15. Change $3\frac{9}{11}$ to an improper fraction
16. $4\frac{2}{3} - 2\frac{1}{2} =$
17. $1\frac{1}{2} \times 5\frac{2}{3} =$
18. $6\frac{1}{4} \div 1\frac{1}{4} =$
19. $2\frac{7}{8} + 3\frac{5}{8} =$
20. $6 \div \frac{3}{4} =$

Check your work and then do the Unit Three Test without your notes.

UNIT THREE **Test**

Set your work out neatly on lined paper. Put your name at the top.

1. What fraction does this represent?

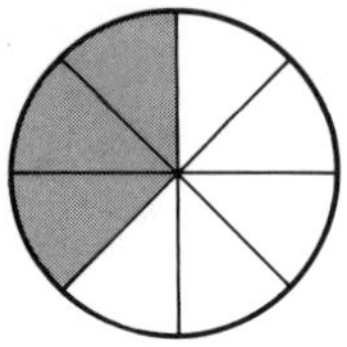

2. Draw a diagram to represent $\frac{3}{5}$

3. Simplify $\frac{15}{20}$

4. Change $1\frac{1}{5}$ to fifths

5. Simplify $\frac{18}{3}$

6. $2\frac{1}{5} + 1\frac{1}{3} =$

7. $1\frac{1}{3} - \frac{5}{6} =$

8. $3\frac{1}{4} \times 1\frac{1}{3} =$

9. $2\frac{1}{4} \div 3 =$

10. Find $\frac{3}{8}$ of 64

Do not forget to check your work.

UNIT THREE ANSWERS

A $\frac{1}{4}$ $\frac{3}{8}$ $\frac{2}{5}$ $\frac{1}{6}$ $\frac{5}{12}$ $\frac{1}{3}$

$\frac{7}{8} =$ 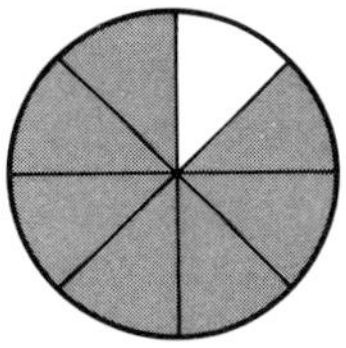 $\frac{3}{5} =$ 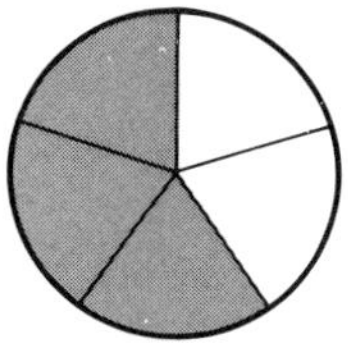

$\frac{4}{5} =$ 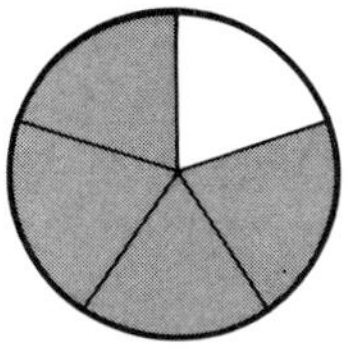 $\frac{7}{12} =$ 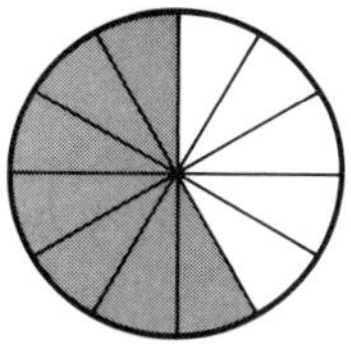

$\frac{2}{3} =$ 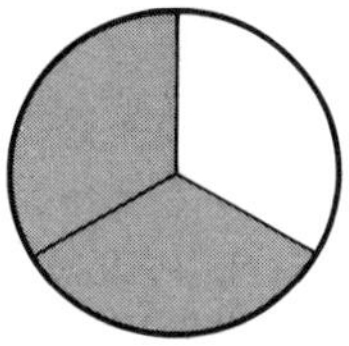 $\frac{2}{9} =$ 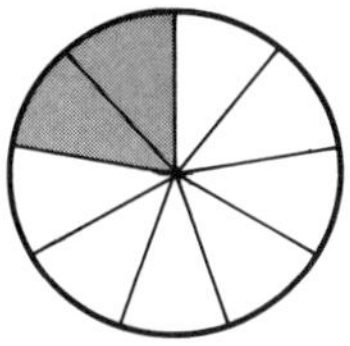

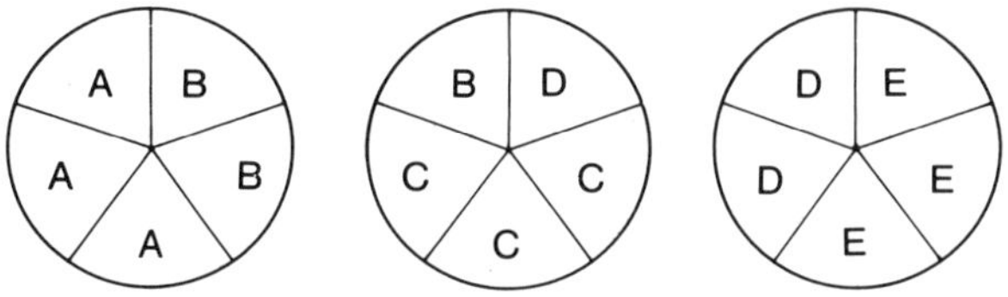

B Exercise 1

1. $\frac{4}{5}$ 2. $\frac{1}{6}$ 3. $\frac{1}{3}$

4. $\frac{3}{4}$ 5. $\frac{2}{3}$ 6. $\frac{1}{6}$

7. $\frac{1}{3}$ 8. $\frac{4}{5}$ 9. $\frac{6}{7}$

10. $\frac{5}{8}$

Exercise 2

1. $\frac{6}{8}$ 2. $\frac{6}{12}$ 3. $\frac{15}{20}$

4. $\frac{6}{18}$ 5. $\frac{15}{24}$ 6. $\frac{44}{80}$

Exercise 3

1. $\frac{5}{2}$ 2. $\frac{25}{4}$ 3. $\frac{46}{9}$

4. $\frac{11}{6}$ 5. $\frac{8}{3}$ 6. $\frac{34}{3}$

7. $\frac{109}{9}$ 8. $\frac{23}{4}$ 9. $\frac{27}{8}$

10. $\frac{30}{7}$

Exercise 4

1. $2\frac{2}{9}$ 2. $4\frac{1}{4}$ 3. 4

4. $6\frac{1}{3}$ 5. $2\frac{1}{2}$ 6. $3\frac{3}{4}$

7. $4\frac{1}{5}$ 8. $4\frac{4}{7}$ 9. 2

10. $13\frac{1}{2}$

C $1\frac{1}{2} + 1\frac{1}{4} = 2\frac{2}{4} + \frac{1}{4} = \mathbf{2\frac{3}{4}}$

$5\frac{3}{4} + 5\frac{2}{3} = 8\frac{9}{12} + \frac{8}{12} = 8\frac{17}{12} = \mathbf{9\frac{5}{12}}$

1. $2\frac{5}{6}$ 2. $3\frac{3}{10}$ 3. $4\frac{2}{9}$

4. $5\frac{9}{20}$ 5. $6\frac{5}{12}$ 6. $1\frac{43}{60}$

7. $2\frac{13}{16}$ 8. $40\frac{5}{8}$ 9. $330\frac{1}{6}$

10. $198\frac{24}{35}$

D $5\frac{1}{2} - 1\frac{1}{3} = 4\frac{3-2}{6} = \mathbf{4\frac{1}{6}}$

$4\frac{1}{4} - 2\frac{2}{3} = 2\frac{3-8}{12} = 1\frac{12+3-8}{12}$

$= \mathbf{1\frac{7}{12}}$

1. $1\frac{1}{6}$ 2. $2\frac{1}{8}$ 3. $\frac{5}{12}$

4. $2\frac{1}{3}$ 5. $2\frac{5}{6}$ 6. $1\frac{5}{6}$

7. $10\frac{13}{18}$ 8. $2\frac{7}{24}$ 9. $57\frac{1}{8}$

10. $58\frac{17}{36}$

E

$$1\tfrac{1}{2} \times 1\tfrac{1}{3} = \frac{^{1}\cancel{3}}{_{1}\cancel{2}} \times \frac{\cancel{4}^{2}}{\cancel{3}_{1}} = 2$$

$$1\tfrac{3}{10} \times \tfrac{5}{8} = \frac{13}{_{2}\cancel{10}} \times \frac{\cancel{5}^{1}}{8} = \mathbf{\frac{13}{16}}$$

1. $4\frac{1}{2}$ 2. $3\frac{3}{8}$ 3. 2

4. $\frac{2}{5}$ 5. $\frac{9}{10}$ 6. $4\frac{4}{9}$

7. $2\frac{1}{16}$ 8. $4\frac{1}{3}$ 9. $6\frac{3}{8}$

10. 2

F

$$1\tfrac{1}{2} \div 2\tfrac{1}{4} = \tfrac{3}{2} \div \tfrac{9}{4} = \frac{^{1}\cancel{3}}{_{1}\cancel{2}} \times \frac{\cancel{4}^{2}}{\cancel{9}_{3}} = \mathbf{\frac{2}{3}}$$

$$3\tfrac{1}{7} \div 2\tfrac{5}{14} = \tfrac{22}{7} \div \tfrac{33}{14} = \frac{^{2}\cancel{22}}{_{1}\cancel{7}} \times \frac{\cancel{14}^{2}}{\cancel{33}_{3}} = \frac{4}{3}$$

$$= \mathbf{1\tfrac{1}{3}}$$

1. $\frac{1}{2}$ 2. 2 3. $2\frac{1}{2}$

4. $\frac{3}{4}$ 5. $6\frac{7}{8}$ 6. $3\frac{1}{4}$

7. $1\frac{1}{3}$ 8. $\frac{1}{5}$ 9. 12

10. $6\frac{13}{18}$

G

$$0.65 = \frac{65}{100} = \frac{13}{20}$$

$$2.175 = 2\frac{175}{1000} = 2\frac{7}{40}$$

1. $\frac{1}{2}$ 2. $\frac{1}{4}$ 3. $3\frac{1}{8}$

4. $5\frac{4}{5}$ 5. $17\frac{1}{5}$ 6. $3\frac{5}{8}$

7. $9\frac{3}{20}$ 8. $15\frac{9}{25}$ 9. $121\frac{3}{4}$

10. $100\frac{7}{10}$

H

$$\frac{4}{5} = \mathbf{0.8} \qquad 5\overline{)4.0} \text{ (quotient } \mathbf{0.8}\text{)}$$

I

1. $5\frac{7}{10}$ 2. $\frac{11}{12}$ 3. $\frac{1}{6}$

4. 0.75 5. $2\frac{1}{2}$ 6. $\frac{13}{20}$

7. $\frac{9}{12}$ 8. $\frac{1}{4}$ 9. 12

10. $\frac{1}{5}$ 11. $\frac{47}{55}$ 12. $\frac{5}{6}$

13. $1\frac{2}{3}$ 14. 0.375 15. $\frac{42}{11}$

16. $2\frac{1}{6}$ 17. $8\frac{1}{2}$ 18. 5

19. $6\frac{1}{2}$ 20. 8

UNIT FOUR

Money

In the United Kingdom we use pounds (£) and pence (p) as units of money. There are 100 pence in a pound. Other countries use different units but usually use a decimal system.

In France they use francs and there are 100 centimes in a franc.

In Germany they use marks and there are 100 pfennigs in a mark.

In Spain they use pesetas. A peseta is very small so they do not need a smaller unit.

Can you find out what currency they use in the United States of America and Italy?

A Changing Pence to Pounds and Pounds to Pence

There are 100 pence in a pound.
To change pounds to pence, multiply by 100.

Example

£1.62 = 162 p
£2.05 = 205 p

To change pence to pounds, divide by 100.

Example

265 p = £2.65
36 p = £0.36
7 p = £0.07

Note that 7 p is £0.07

Always state the units in the answer: pounds or pence. It can *never* be both, £2.62 p is *wrong.*

Fill in these examples:

Change to pence:	*Change to pounds:*
£7.62 =	149 p =
£4.28 =	68 p =
£0.65 =	1440 p =
£144 =	89 p =
£9.02 =	6 p =

B Addition and Subtraction

Work in either pence or pounds. Keep the decimal points in line.

Example

Add £2.68 + 69 p + £5.26.

£	*Change 69 p to £0.69.*
2.68+	*Write the numbers under each other*
0.69	*with the points in line.*
5.26	*Then add.*
8.63	ANS = **£8.63**
1 2	

Example

Subtract £1.62 from £3.59.

£		£
3.59 −	*or*	3.59 −
1.62		1.62
1.97	ANS = £1.97	1.97

Check by adding £1.62 + £1.97: you should get £3.59.

Example

A man buys a book for £2.99. How much change will he get from a £5 note?

£		£
5.00 −	*or*	5.00 −
2.99		2.99
2.01	ANS = £2.01	2.01

Example

An American boy buys a book for $3.95, and an ice-cream for 95 cents. He gets $5 pocket money. How much has he got left?

$	
3.95+	*95 cents is $0.95*
0.95	*Add the amounts he spends.*
4.90	

$	
5.00−	*Subtract from the amount he starts with.*
4.90	
0.10	

When you are given a problem answer in a sentence.
Always state the units. $0.10 is also a correct answer, but just 10 or 0.10 is wrong because it does not say if it is cents or dollars.

He has *10 cents* left ANS

Fill in these examples:

£3.65 + £1.95 + 55 p

£
3.65+
1.95
0.55

ANS = £

£5.00 − £2.79

£
5.00−
2.79

ANS = £

If you get them right try these. Set them out neatly on paper and don't forget the units:

1. £2.54 + £9.29 =
2. £11.72 + £4.63 =
3. £104.79 + £9.86 + £5.74 =
4. £6.55 − £3.24 =
5. £18.16 − £9.07 =
6. £1.00 − 67 p
7. 6.72 marks + 3.68 marks =
8. A girl has 11 francs. She spends 2 francs and 35 cents. How much has she left?
9. If a movie costs $4.95, a hamburger $1.95 and a coke 75 cents, how much does it cost a boy and girl to have an evening out at the movies with a hamburger and coke each?
10. A man buys lunch in his works canteen each day. On Monday his lunch costs 95 p, Tuesday £1.05, Wednesday £1.25, Thursday 65 p and Friday 86 p. He gets £2.00 worth of luncheon vouchers towards his meals. How much money does he spend for the week's meals?

Check your work before continuing.

C Multiplication

To multiply money is exactly like decimal multiplication.

Example

```
  £
 5.65×
    6
 33.90
  3 3
```

Ignore the decimal point and multiply by 6. There are two numbers after the decimal point in the question so it goes two places from the right in the answer.

ANS = **£33.90**

Example

```
  £
 3.95×
    6
 23.70
  5 3
```

ANS = **£23.70**

Example

A woman buys $4\frac{1}{2}$ kg of seed at 67 p a kilogram. Find how much she pays (answer to nearest penny below):

67 p = £0.67 *Change both to decimals.*

$4\frac{1}{2}$ kg = 4.5 kg

```
 0.67×
  4.5
  335
 2680
3.015
```

Multiply 0.67 × 4.5 and you get 3.015. She pays £3.015. This is £3.02 to the nearest penny.

She pays **£3.02** ANS

When we multiply and divide money we often get answers with more than 2 figures after the decimal point e.g. £2.6147. This is £2 and 61 pence and part of a penny. The rule is if the part of a penny starts with 5 or more add 1 to the pence, otherwise ignore it.

Thus £2.6147 becomes £2.61
£27.0693 becomes £27.07
£18.398 becomes £18.40

Write to the nearest penny:

1. £3.75 =
2. £6.968 =
3. £8.642 =
4. Multiply £6.94 × 5

```
  £
 6.94×
    5
 ____
```

ANS =

5. Multiply £5.82 × 7

```
  £
 5.82×
    7
 ____
```

ANS =

If you did these correctly you can try these on paper. *Don't forget the units. Work neatly in columns:*

1. £2.67 × 8
2. £105.62 × 3
3. $4.96 × 5
4. £3.82 × $2\frac{1}{2}$
5. £6.98 × 12
6. 68 p × 6
7. 15 p × 8
8. 2 p × 17
9. Find the cost of 8 shirts at £14.99 each.
10. Find the cost of $3\frac{1}{2}$ m of curtain material at £3.99 a metre.

D Division

To divide money use the same method as decimals. You will only need to divide by whole numbers.

Example

Share £5.25 between five people, so they will have the same amount.

$$\begin{array}{r} \pounds \\ 1.05 \\ 5\overline{)5.25} \end{array}$$

Each person gets **£1.05** ANS

Example

Share £3.28 equally between four people.

$$\begin{array}{r} \pounds \\ 0.82 \\ 4\overline{)3.28} \end{array}$$

Each person gets **82** p ANS

Fill in these answers:
£6.95 ÷ 5

$$\begin{array}{r} \pounds \\ \\ 5\overline{)6.95} \end{array}$$

ANS =

£8.62 ÷ 3
(to the nearest penny)

$$\begin{array}{r} \pounds \\ \\ 3\overline{)8.620} \end{array}$$

ANS =

Then try these on paper. All answers to the nearest penny or cent:

1. £8.59 ÷ 3 =
2. £17.41 ÷ 6 =
3. $5.00 ÷ 4 =
4. £5.92 ÷ 7 =
5. £102.15 ÷ 3 =

E Foreign Exchange

When you go abroad you have to change your money into foreign currency. The amount you get varies each day and is listed in banks and some newspapers.
It may look something like this:

TOURIST RATES	
AUSTRIA	26.00 schillings
BELGIUM	75.50 francs
FRANCE	10.58 francs
WEST GERMANY	3.73 marks
ITALY	2130 lire
SPAIN	195 pesetas
UNITED STATES	1.56 dollars

This tells you how much foreign currency you will get for each pound.

Example

Change £100 to dollars:

£1 = $1.56 — *Write down the conversion rate.*

£100 = $1.56 × 100 — *£100 will give 100 times as much as £1 so multiply by 100*

= $156

Example

Change £50 to marks:

Write down the conversion rate.

£1 = 3.73 marks

£50 will give 50 times as much so multiply by 50.

£50 = 3.73 × 50 marks

$$\begin{array}{r} \mathbf{3.73\times} \\ \mathbf{50} \\ \hline 186.50 \\ \hline \end{array}$$

= 186.5 marks

Fill in the answers using the rates opposite:

1. Change £100 to French francs:

£1 =

£100 = ANS

2. Change £20 to lire:

£1 =

£20 = ANS

3. Change £200 to pesetas:

£1 =

£200 = ANS

4. Change £50 to Austrian currency:

£1 =

£50 = ANS

5. Change £40 to Belgian francs:

£1 =

£40 = ANS

If you have some money left after your holiday you will need to change it back to British currency. Given the conversion rate from pounds to marks you will have to turn it round to get a conversion from marks to pounds, like this:

£1 = 4 marks

then 1 mark = $£\frac{1}{4}$

so to convert your marks to pounds you divide by 4.

$$\text{20 marks} = £\frac{20}{4} = £5$$

$$\text{100 marks} = £\frac{100}{4} = £25$$

You divide because you are using the conversion the other way round

pounds → marks
multiply

£1 = 4 marks

marks → pounds
divide

Example

Change 520 schillings to pounds:

£1 = 26 schillings

$$\text{1 schilling} = £\frac{1}{26}$$

$$\text{520 schillings} = £\frac{520}{26}$$

$$= £20$$

Try these:

6. Change $156 to pounds:

£1 = $

$$\$1 = £\frac{1}{\ }$$

$$\$156 = £\frac{156}{\ }$$

$$= £ \quad \text{ANS}$$

7. Change 975 pesetas to pounds:

£1 = pesetas

$$\text{1 peseta} = £\frac{1}{\ }$$

$$\text{975 pesetas} = £\frac{975}{\ }$$

$$= £ \quad \text{ANS}$$

8. Change 74.6 marks to pounds:

£1 = marks

$$\text{1 mark} = £\frac{1}{\ }$$

$$\text{74.6 marks} = £\frac{74.6}{\ }$$

$$= £ \quad \text{ANS}$$

F More Conversion

When you have to convert currency always write down the conversion rate first. Then decide if you need to multiply or divide. (Is it more or less?) Then work out the calculation. If the problem has several parts tackle one at a time. You *must* show all your working.

Example

A girl goes to France on holiday. She takes £100 with her. She spends 920 francs and converts the remainder back to pounds. Given that £1 = 11 francs, how many pounds does she have left?

£1 = 11 francs *Write down the conversion rate.*

£100 = 100 × 11 francs *She gets more francs so multiply.*

= 1100 francs

1100 −
920 *Take away the amount she spends.*

180 francs. *She has 180 francs left.*

£1 = 11 francs

1 franc $= £\frac{1}{11}$

180 francs $= £\frac{180}{11}$ *Convert back to pounds. She gets less so divide.*

$11\overline{)180.000}$ (carries: 7 4 7 4)

16.363

= £16.36 (nearest penny)

She has £16.36 left ANS

You fill in this one:

A girl goes to France with £70. She spends 400 francs. How many pounds does she have left? Answer to the nearest pound. (Given £1 = 11 francs.)

£1 = 11 francs

£70 = 70 × 11 francs

= francs

−

400 *subtract.*

£1 = 11 francs

1 franc $= £\frac{1}{11}$

francs $= £\frac{\quad}{11}$

= £

She has £ left ANS

If you got that problem right you will be able to tackle these on paper. Think carefully about whether you must multiply or divide:

1. Change £100 to US dollars (£1 = \$1.54).
2. Change 577.50 francs to pounds (£1 = 10.5 francs).
3. Change 144.4 marks to pounds (£1 = 3.8 marks).
4. Change £120 to lire (£1 = 2200 lire).
5. A man takes £200 to Germany. He spends 500 marks. He changes the remainder to pounds. How much does he have left? (£1 = 3.8 marks.)
6. A train ticket costs 550 francs. I buy it in England. How much do I pay? (£1 = 10.5 francs.)

G Mixed Questions on Money

Try these questions on paper. Refer back to your previous work if you need to:

1. Change 795 p to pounds.
2. Change £12.48 to pence.
3. £5.96 + £2.88 + 96 p =
4. £8.15 − £2.19 =
5. I buy three stamps at 17 p each. How much change do I get from £1?
6. $8.65 × 4 =
7. 9.24 marks ÷ 6 =
8. Find the cost of 15 French cakes at 3.50 francs each.
9. Share £210 equally between eight people.
10. Change £150 to marks (£1 = 3.8 marks).
11. Change 200 francs to pounds (£1 = 11.2 francs) (to nearest penny).
12. A family go on holiday with £500 spending money. They spend ten days in France and four days in Germany. In total they spend 3150 francs and 408 marks. How much have they left after their holiday? (£1 = 10.8 francs, £1 = 3.9 marks.) Work each conversion to the nearest penny.

When you have corrected your work you are ready to do Unit Four Test.

UNIT FOUR **Test**

Set your work out neatly on lined paper. Put your name at the top.

1. Change £27.48 to pence.
2. Change 6098.54 p to pounds. (Answer to the nearest penny.)
3. £6.87 + £96 + 2 p
4. £10 − £9.67
5. Find the change from £10 when I buy $2\frac{1}{2}$ metres of material at £1.99 a metre.
6. Change £900 to dollars (£ = $1.54).
7. Change 2730 francs to pounds (£1 = 10.5 francs).
8. In France a man buys four garden chairs for a total of 112 francs. How much are they each? What is the equivalent price each in pounds? (£1 = 10.5 francs.) Answer to nearest penny.
9. £2.54 × 17 =
10. Share 89 p equally between six people to the nearest penny.

Do not forget to check your work.

In America they use dollars and cents.
In Italy they use lire.

A

762 p	£1.49
428 p	£0.68
65 p	£14.40
14 400 p	£0.89
902 p	£0.06

B

$$\begin{array}{r} \pounds \\ 3.65+ \\ 1.95 \\ 0.55 \\ \hline \pounds 6.15 \\ \hline \end{array}$$

ANS = £6.15

$$\begin{array}{r} \pounds \\ {}^{4}\not{5}.{}^{9}\not{0}{}^{1}0- \\ 2.79 \\ \hline 2.21 \\ \hline \end{array} \qquad \textit{or} \qquad \begin{array}{r} \pounds \\ 5.{}^{1}0{}^{1}0- \\ {}^{1}2.{}^{1}79 \\ \hline \pounds 2.21 \\ \hline \end{array}$$

ANS = £2.21

1. £11.83
2. £16.35
3. £120.39
4. £3.31
5. £9.09
6. 33 p or £0.33
7. 10.40 marks
8. 8.65 francs
9. $15.30
10. £2.76

C

1. £3.75
2. £6.97
3. £8.64
4.

$$\begin{array}{r} 6.94\times \\ 5 \\ \hline 34.70 \\ \hline {}_{4\ 2} \end{array}$$

ANS = £34.70

5.

$$\begin{array}{r} 5.82\times \\ 7 \\ \hline 40.74 \\ \hline {}_{5\ 1} \end{array}$$

ANS = £40.74

1. £21.36
2. £316.86
3. $24.8
4. £9.55
5. £83.76
6. £4.08
7. £1.20
8. 34 p
9. £119.92
10. £13.97

D

$$\begin{array}{r} 5\overline{)6.{}^{1}9{}^{4}5} \\ 1.39 \\ \hline \end{array}$$

ANS = £1.39

$$\begin{array}{r} 3\overline{)8.{}^{2}6{}^{2}2{}^{1}0} \\ 2.873 \\ \hline \end{array}$$

ANS = £2.87

1. £2.86
2. £2.90
3. $1.25
4. 85 p
5. £34.05

E

1. 1058 French francs
2. 42 600 lire
3. 39 000 pesetas
4. 1300 schillings
5. 3020 Belgian francs

6. $\pounds 1 = \$1.56$

$\$1 = \pounds\frac{1}{1.56}$

$\$156 = \pounds\frac{156}{1.56} = \pounds 100$

7. **£1 = 195 pesetas**

1 peseta $= \pounds\frac{1}{195}$

975 pesetas $= \pounds\frac{975}{195}$

$= \pounds 5$

8. **£1 = 3.73 marks**

74.6 marks

$= \pounds\frac{74.6}{3.73}$

$= \pounds 7460 \div 373$

$= \pounds 20$

$$\begin{array}{r} 20 \\ 373\overline{)7460} \\ 746 \\ \hline 0000 \end{array}$$

F

$$
\begin{aligned}
£1 &= 11 \text{ francs} \\
£70 &= 70 \times 11 \text{ francs} \\
&= 770 \text{ francs}
\end{aligned}
$$

$$
\begin{array}{r}
770\ - \\
\underline{400} \\
\underline{370}
\end{array}
$$

Subtract.

$$
\begin{aligned}
£1 &= 11 \text{ francs} \\
1 \text{ franc} &= £\frac{1}{11} \\
370 \text{ francs} &= £\frac{370}{11} \\
&= £33.6
\end{aligned}
$$

She has £34 left (to nearest pound) ANS

1. $154
2. £55
3. £38
4. 264 000 lire
5. He starts with 760 marks. He has 260 marks left which gives him **£68** (£68.42).
6. £52.38

G

1. £7.95
2. 1248 p
3. £9.80
4. £5.96
5. 49 p
6. $34.6
7. 1.54 marks
8. 52.50 francs
9. £26.25
10. 570 marks
11. £17.86
12. 3150 francs = £291.67
    408 marks = £104.62

    They have **£103.71** left.

UNIT FIVE

Weights and Measures

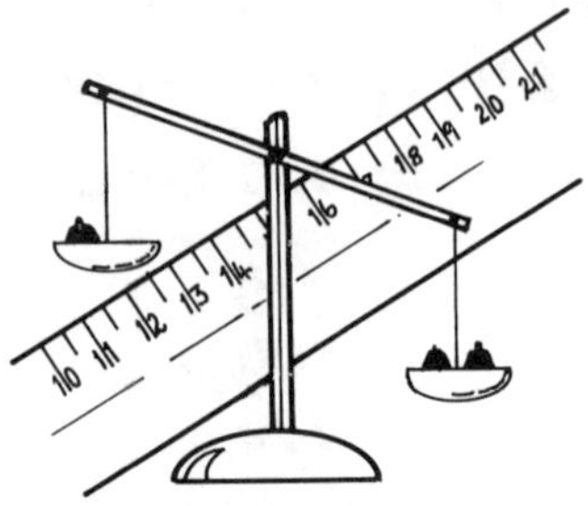

A Metric Measures

The metric system is based on tens.

Length The most common units are:

10 millimetres (mm)	=	1 centimetre (cm)
100 centimetres	=	1 metre (m)
1000 metres	=	1 kilometre (km)
1000 millimetres	=	1 metre

This line is 4.3 cm long. That is the same as 43 mm.

Check it on your ruler.

Now measure this line.

Write down the length in centimetres.
ANS = cm

Write it down in millimetres.
ANS = mm

Weight

1000 grams (g) = 1 kilogram (kg)
1000 kilograms = 1 tonne

Capacity

10 millilitres (mℓ) = 1 centilitre (cℓ)
1000 millilitres = 1 litre (ℓ)
100 centilitres = 1 litre

You need to learn these tables

Read them through several times then cover them up and try to fill in the gaps from memory:

1.	m = 1 km	**6.**	10 mm = 1
2.	mm = 1 m	**7.**	mℓ = 1 cℓ
3.	1 kg = g	**8.**	100 = 1 ℓ
4.	1000 mℓ = 1	**9.**	1000 g = 1
5.	cm = 1 m	**10.**	1 m = 100

If you found it difficult to learn the tables try writing them out several times then attempt the test again. The later work will be a lot easier if you know the tables.

B Imperial Measures

A lot of people in Britain (some shops, milkmen etc.) still use imperial measures so you have to know these as well.

Length

12 inches (in)	=	1 foot (ft)
3 feet	=	1 yard (yd)
1760 yards	=	1 mile

This line is $2\frac{3}{8}$ in long.

Rulers marked in inches are usually divided into 8 or 16 parts.

Measure this line in inches. ANS = in

Weight

16 ounces (oz) = 1 pound (lb)

Other units include stones, hundredweight, tons. You need not learn them by heart but they are:

14 lb = 1 stone
112 lb = 8 stone = 1 hundredweight (cwt)
20 cwt = 2240 lb = 1 ton

Capacity

8 pints (pt) = 1 gallon (gal)

Now try to fill in the gaps from memory:

1.	oz = 1 lb
2.	in = 1 ft
3.	yd = 1 mile
4.	pt = 1 gal
5.	ft = 1 yd

C Changing Units

The same quantity can be expressed in various ways.

Example

4.3 cm = 43 mm = 4 cm 3 mm (cm and mm is 'mixed units').

This is because each cm = 10 mm.

Fill in the gaps (look back to the tables if you need to):

1. 3.6 cm =	mm =	cm	mm		
2. 4.5 kg =	g =	kg	g		
3. 5.2 m =	cm =	m	cm		
4. 620 cm =	m =	m	cm		
5. 6.258 m =	cm =	mm =	m	cm	mm
6. 26 g =	kg				
7. 1450 g =	kg =	kg	g		
8. 2.48 ℓ =	mℓ =	ℓ	mℓ		
9. 524 cℓ =	ℓ =	ℓ	cℓ		
10. 1640 m =	km =	km	m		

When you have got used to the metric units try the imperial ones (be careful!). The first one is done for you:

1. 18 in =	1 ft 6 in			
2. 25 in =	ft	in		
3. 59 in =	yd	ft	in	
4. 2760 yd =	mile	yd		
5. 49 oz =	lb	oz		
6. $2\frac{1}{2}$ lb =	lb	oz =	oz	
7. 26 pt =	gal	pt		
8. $2\frac{1}{4}$ gal =	gal	pt =	pt	
9. 2 yd 1 ft =	ft =	in		
10. 33 lb =	stone	lb	(14 lb = 1 stone)	

If you got all these right then well done!

D Calculations in Metric Units

When you are adding and subtracting convert all the quantities to the same unit.

Example

36 m + 172 cm + 3 m 15 cm

```
 3600+      Change everything to cm.
  172
  315
 ----
 4087       ANS = 4087 cm,
```

or we could do the same sum in metres:

```
 36    +
  1.72
  3.15
 -----
 40.87      ANS = 40.87 m
```

You can choose which units you work in. If you are asked for the answer in centimetres, it is 4087 cm, in mixed units it is 40 m 87 cm, or in metres 40.87 m.

Example

A piece of string is 2.58 m long. If 92 cm is cut off, how much is left? (Answer in centimetres.)

$$\begin{array}{r} 258- \\ 92 \\ \hline 166 \\ \hline \end{array}$$

Change 2.58 m to centimetres.
Subtract.

166 cm are left ANS

Answer problems with a sentence.

Fill in these examples:

1. A shop has 60 kg of apples. Three customers buy 2.5 kg, 600 g and $1\frac{1}{2}$ kg. How many kilograms are left?

   2.5+   60−
   0.6
   1.5

   There are ____ kg left ANS

2. A hotel chef mixes 2.3 kg tomatoes, 800 g cucumber, 1.2 kg lettuce and 65 g of radishes to make a salad.
   How many grams of salad does she make?

   2300+

   She makes ____ g of salad ANS

Write out these examples on paper:

3. Add together 5 km + 6.92 km + 900 m.

4. 2.96 ℓ − 315 mℓ =

5. A girl has 5 m of cord. She uses 80 cm in packing each of two parcels. For a larger package she uses 1.2 m, and to hang a picture 45 cm. How much is left?

E Multiplication and Division

The method is just like decimals. Take care not to make careless mistakes.

Example

Five girls are to be dressed alike in a play. Each costume takes 2.4 m of material. How much is needed altogether?

$$\begin{array}{r} 2.4\times \\ 5 \\ \hline 12.0 \\ \hline \end{array}$$

12 m of material is needed ANS

Example

A dairy has 40 kg of yoghurt. How many 125 g pots can they fill?

40 kg = 40 000 g   *Change to grams.*

$$\begin{array}{r} 320 \\ 125\overline{)40\,000} \\ 375 \\ \hline 250 \\ 250 \\ \hline 0 \end{array}$$

Divide.

They can fill **320 pots** ANS

Fill in these:

1. The frame of a cube needs 12 lengths of wood each 69 cm long. Find the total length in metres.

   69 ×
   12
   ——

   ——
   cm = m The length is m ANS
   ——

2. A cafe has 200 ℓ of lemonade. If each glass holds 300 mℓ, how many glasses can they fill?

   $300\overline{)}$ They fill glasses ANS

   Answer these on paper:

3. Find the total weight of 12 cartons weighing 600 g and 24 cartons weighing 250 g. (Answer in kilograms and grams.)

4. A baker uses 1.4 kg of icing sugar for each large cake and 0.8 kg for a small one. He is going to ice five large cakes and eight small. How much icing sugar will he need?

5. Three brothers are given 1 kg of sweets for Christmas. How much will they get each? (Answer to the nearest gram.)

F Imperial Calculations

Example

2 lb 8 oz + 1 lb 11 oz + 4 oz

lb	oz	
2	8	*8 + 11 + 4 = 23 oz = 1 lb 7 oz*
1	11	
	4	
4	7	ANS = **4 lb 7 oz**

Example

A string 2 ft 4 in long is cut into four equal parts. How long is each part?

2 ft 4 in = 28 in

$$4\overline{)28}\quad = 7$$

Each part is **7 in** long ANS

Try these on paper:

1. 3 ft 6 in + 2 ft 5 in + 7 in =
2. Four girls wish to run a total of one mile in a relay race. How many yards must they run each?
3. Share 2 lb of sweets equally between eight children.
4. How many $\frac{1}{2}$ pt mugs can be filled from 8 gal?
5. 6 gal − 4 gal 2 pt =
6. 2 ft 3 in × 9 = (Answer in feet and inches.)
7. 1 yd ÷ 9 = (Answer in inches.)
8. 1 lb 2 oz + $4\frac{1}{2}$ oz + 2 lb $9\frac{1}{2}$ oz =
9. The weight of beans picked each day in the garden was 1 lb 2 oz, 10 oz, 2 lb 4 oz, 15 oz, 1 lb 1 oz, 14 oz, $8\frac{1}{2}$ oz. Find the total for the week.
10. 105 infants each get $\frac{1}{3}$ pt of milk a day. Milk costs 20 p a pint. How much will the milk cost for a week (five days)?

G Changing Between Metric and Imperial Units

Some petrol stations sell petrol in gallons, others in litres. Some foodstuffs are packed in ounces, others in grams.
In order to compare sizes we need to change from one measuring system to the other.

Some useful conversions are:

1 in = 2.54 cm	1 cm = 0.39 in
1 lb = 0.45 kg	1 kg = 2.2 lb
1 mile = 1.6 km	1 km = 0.62 mile
1ℓ = 0.22 gal	1 gal = 4.5 ℓ

These are approximate conversions.
Do not learn them off by heart. When you need them you can look them up (they will be given in an exam).

Example

Change 6 in to centimetres.

1 in = 2.54 cm *Write down the conversion for one.*

6 in = 6 × 2.54 cm *Multiply.*

= **15.24 cm**

Note that if the conversion is not the required way round then use this method.

Example

Change 27 ℓ to gallons (Given 1 gal = 4.5 ℓ).

1 gal = 4.5 ℓ

$1\,\ell = \frac{1}{4.5}$ gal *Change the conversion round.*

$27\,\ell = 27 \times \frac{1}{4.5}$ gal *Multiply by 27.*

$= \frac{27}{4.5} = \frac{270}{45}$ *This gives a division.*

= **6 gal**

$$\begin{array}{r} 6 \\ 45\overline{)270} \\ 270 \\ \hline 0 \end{array}$$

Try these first. Choose the conversion you need from the list above:

1. Change 5 in to centimetres.
2. Change 8 lb to kilograms.
3. Change 7 kg to pounds.
4. Change 50 miles to kilometres.
5. Change 8 gal to litres.

When you have checked these from the answers, try these using the conversion given. Change it round if necessary:

6. Change 7 miles to kilometres. (Use 1 mile = 0.621 km.)
7. Change 54 ℓ to gallons. (Use 1 gal = 4.5 ℓ.)
8. Change 1 ft 9 in to centimetres. (Use 1 in = 2.54 cm.)
9. Change 95 cm to feet and inches. (Use 1 in = 2.5 cm.)
10. Change 100 kg to pounds. (Use 1 kg = 2.205 lb.)

H Mixed Questions on Weights and Measures

Now try some mixed examples. Look back through the notes if you get stuck:

1. How many grams in a kilogram?
2. How many ounces in a pound?
3. Change 540 cm to metres and centimetres.
4. Write 2.935 kg in grams.
5. Change $5\frac{1}{2}$ gal to pints.
6. Find the total length of five pieces of string each 1 ft $3\frac{1}{2}$ in long.
7. Share 50 kg of coal equally between 16 people.
8. 2 m 63 cm − 1 m 18 cm =
9. A tank holds 600 ℓ of milk. How many pint bottles will it fill? (Use 1 ℓ = 0.22 gal.)
10. 3 m 18 cm × 10 (Answer in metres and centimetres.)
11. 19 ℓ ÷ 5 = (Answer in centilitres.)
12. A gallon of petrol costs £1.68. What is the price per litre? (Use 1 gal = 4.5 ℓ.) (Answer to the nearest penny.)

Now go on to Unit Five Test.

UNIT FIVE **Test**

Set your work out neatly on lined paper. Put your name at the top.

1. Change 6.92 cm to millimetres.
2. Express 8 kg 16 g in kilograms.
3. Change 2 yd 1 ft to feet.
4. $6\,\ell \div 8 =$ (Answer in millimetres.)
5. 6.52 kg + 914 g + 4.6 kg =
6. 2 ft − 11 in =
7. How many coins each 2 cm in diameter placed end to end will cover a kilometre?
8. Convert 2 ft 2 in to metres (1 in = 2.54 cm).
9. Convert $90\,\ell$ to gallons (a gal = $4.5\,\ell$).
10. A piece of material is 9 m long. If I use 2.4 m for a dress, 1.8 m for a skirt and $3\frac{1}{2}$ m for a shirt how much will be left?

Do not forget to check your answers.

UNIT FIVE ANSWERS

A 6.7 cm 67 mm

1.	1000	**2.**	1000	**3.**	1000
4.	ℓ	**5.**	100	**6.**	cm
7.	10	**8.**	cℓ	**9.**	kg
10.	cm				

B $1\frac{3}{4}$ in

1.	16	**2.**	12	**3.**	1760
4.	8	**5.**	3		

C

1. 3.6 cm = 36 mm = 3 cm 6 mm
2. 4.5 kg = 4500 g = 4 kg 500 g
3. 5.2 m = 520 cm = 5 m 20 cm
4. 620 cm = 6.2 m = 6 m 20 cm
5. 6.258 m = 625.8 cm = 6258 mm = 6 m 25 cm 8 mm
6. 26 g = 0.026 kg
7. 1450 g = 1.45 kg = 1 kg 450 g
8. 2.48 ℓ = 2480 mℓ = 2 ℓ 480 mℓ
9. 524 cℓ = 5.24 ℓ = 5 ℓ 24 cℓ
10. 1640 m = 1.64 km = 1 km 640 m

2. 2 ft 1 in
3. 1 yd 1 ft 11 in
4. 1 mile 1000 yd
5. 3 lb 1 oz
6. 2 lb 8 oz = 40 oz
7. 3 gal 2 pt
8. 2 gal 2 pt = 18 pt
9. 7 ft = 84 in
10. 2 stone 5 lb

D

1.
$$\begin{array}{r} 2.5\ + \\ 0.6 \\ 1.5 \\ \hline 4.6 \\ \hline \end{array} \qquad \begin{array}{r} 60.0\ - \\ 4.6 \\ \hline 55.4 \\ \hline \end{array}$$

There are **55.4 kg** left.

2.
$$\begin{array}{r} 2300\ + \\ 800 \\ 1200 \\ 65 \\ \hline 4365 \\ \hline \end{array}$$

She makes **4365 g** of salad.

3. 12.82 km
4. 2.645 ℓ or 2645 mℓ
5. 175 cm or 1.75 m

E

1.
$$\begin{array}{r} 69\times \\ 12 \\ \hline 138 \\ 690 \\ \hline 828 \\ \hline \end{array}$$

828 cm = 8.28 m

The length is **8.28 m.**

2.
$$300\overline{)200\,000}\quad 666$$

They fill **666** glasses [NB some is left over].

3. 13 kg 200 g
4. 13.4 kg (or 13 kg 400 g)
5. 333 g

F

1. 6 ft 6 in
2. 440 yd
3. 4 oz or $\frac{1}{4}$ lb
4. 128
5. 1 gal 6 pts
6. 20 ft 3 in
7. 4 in
8. 4 lb
9. 7 lb $6\frac{1}{2}$ oz
10. £35

G

1. 12.7 cm
2. 3.6 kg
3. 15.4 lb
4. 80 km
5. 36 ℓ
6. 4.347 km
7. 12 gal
8. 53.34 cm
9. 3 ft 2 in
10. 220.5 lb

H

1. 1000
2. 16
3. 5 m 40 cm
4. 2935 g
5. 44 pt
6. 6 ft $5\frac{1}{2}$ in
7. 3.125 kg
8. 1 m 45 cm
9. 1056
10. 31 m 80 cm
11. 380 cℓ
12. 37 p

UNIT SIX

Magnitude

Magnitude means 'size'.

A Rounding Money

When we calculate amounts they don't always work out exactly. Then it is better to round them off. For instance, if you bought ten pens for £1.48 and wanted to sell one to your friend you would charge 15 p since you can't give 14.8 p in coins.

Banks always work to the nearest penny. If you are measuring your window to buy curtains you would measure to the nearest inch or centimetre.

If you have £12.98, how many £1 tokens can you buy? The answer is 12. Although £12.98 is nearer £13 than £12 you can only buy 12 tokens. Look out for questions like this.

Imagine you are a bank cashier.
Round off these amounts to the nearest penny (0.5 goes up):

	Customer	*Interest*	*To nearest penny*
a)	Allan	£10.689	£10.69
b)	Brown	£104.695	
c)	Cooper	£28.5449	
d)	Dawson	£5.271	
e)	Eagle	£82.999	
f)	French	£16.79	
g)	Gordon	£100.917	
h)	Henderson	£33.333	
i)	Ivan	£0.0287	
j)	Jones	£17.6892	

Now be a shopkeeper and round these prices to the nearest penny:

	Item	*Calculated price*	*To nearest penny*
k)	Kippers	87.936 p	88 p
l)	Lard	12.64 p	
m)	Marmalade	45.98 p	
n)	Nice Biscuits	21.45 p	
o)	Oranges	12.81 p	
p)	Pineapple Slices	62.08 p	
q)	Quail	£5.8625	
r)	Radishes	28.92 p	
s)	Smarties	13.89 p	
t)	Tomatoes	62.41 p	

B Decimal Places

This method of rounding off a number is also used when working to a given number of decimal places.

Example

Express 3.629 to 2 decimal places.
3.62/9
Count 2 places after the point
= **3.63** (2 decimal places)
If the next digit is 5 or more add one. If less than 5 do nothing.
Express 72.314 to 2 d.p. (decimal places)
72.31/4 = **72.31** (2 d.p.)

Fill in these:

1. 8.6152 to 3 d.p. =
2. 17.118 to 2 d.p. =
3. 1.041 to 1 d.p. =
4. 8.2499 to 2 d.p. =
5. 16.992 to 1 d.p. =

C Significant Figures

We can also round an answer to several significant figures.

Example

Round off 83.28 to 3 sig. fig. (significant figures).

83.2/8

Count 3 figures (ignoring zeros at the beginning.)

= **83.3** (3 sig. fig) *Round off as before.*

Round off 0.001 408 to 3 sig. fig.

0.001 40/8

Don't count the zeros before the 1 but leave them in the answer.

= **0.001 41** (3 sig. fig.)

You do these:

1. 136.293 to 3 sig. fig. =
2. 63.58 to 2 sig. fig. =
3. 0.012 34 to 2 sig. fig. =
4. 0.086 95 to 3 sig. fig. =
5. 1003.8 to 3 sig. fig. =

D Fractions and Order

Can you arrange fractions and decimals in order of size?

Example

Arrange $5\frac{1}{3}$, 5.3 and $5\frac{2}{11}$ in order of size, smallest first.

Change the fractions to decimals, by dividing the top by the bottom.

$\frac{1}{3} = 0.333\ldots$

$\frac{2}{11} = 0.181\,818\ldots$

so, $5\frac{1}{3} = 5.333\ldots$

$5.3 = 5.3$

$5\frac{2}{11} = 5.1818\ldots$

In order of size, smallest first, we have:

$5\frac{2}{11}$, 5.3, $5\frac{1}{3}$

Try these yourself:

1. Arrange in order, biggest first 1.9, $1\frac{8}{9}$, 1.88.
2. Arrange in order, smallest first $\frac{5}{8}$, $\frac{5}{9}$, $\frac{6}{7}$.
3. Arrange in order, biggest first 3.14, 3.139, $3\frac{2}{11}$.

E Squares and Square Roots

To square a number multiply it by itself.

3 squared is written $3^2 = 3 \times 3 = 9$

$$5^2 = 5 \times 5 = 25$$

$$\left(1\tfrac{1}{2}\right)^2 = 1\tfrac{1}{2} \times 1\tfrac{1}{2} = \tfrac{3}{2} \times \tfrac{3}{2} = \tfrac{9}{4} = 2\tfrac{1}{4}$$

Square roots are the reverse operation.

$\sqrt{9} = 3$

$\sqrt{25} = 5$

$\sqrt{2\frac{1}{4}} = 1\frac{1}{2}$

One way of finding a square root is to guess and then see if it works.

$\sqrt{25}$ Guess 5 Try it $5 \times 5 = 25$

Yes it is right $\sqrt{25} = 5$

For more complicated square roots it is necessary to use other methods or a calculator. For this exercise use the 'guessing' method.

(Look for clues in the other questions.)

Fill in the answers:

$\left(2\frac{1}{2}\right)^2$	
25^2	
100^2	
7.1^2	
80^2	

$\sqrt{16}$	
$\sqrt{64}$	
$\sqrt{1\ 000\ 000}$	
$\sqrt{900}$	
$\sqrt{625}$	

F HCF Highest Common Factor

This is the biggest number that will divide into two (or more) others
i.e. the *biggest factor* of both numbers.
A factor is a number that divides into another i.e. 2 and 3 are factors of 6.

Example

Find the HCF of 16 and 24.

The factors of 16 are 1, 2, 4, 8, 16.

(*Notice the factors go in pairs* $1 \times 16 = 16$, $2 \times 8 = 16$, $4 \times 4 = 16$).

The factors of 24 are 1, 2, 3, 4, 6, 8, 12, 24.

Can you see the pairs?

The biggest number in both sets is 8.

The HCF of 16 and 24 is **8** *ANS*

Fill in this example:

Find the HCF of 36 and 60.

The factors of 36 are

The factors of 60 are

(*Check the pairs.*)

The HCF of 36 and 60 is ANS.

Now try these:

1. Find the HCF of 18 and 36.
2. Find the HCF of 6, 9 and 15 (the biggest factor of all three).
3. Find the HCF of 25, 16, 36.
4. Find the HCF of 12, 18, 15.

G LCM Least Common Multiple

A multiple of 3 is 3, 6, 9, 12, etc.
The least common multiple is the smallest multiple of all the numbers.

Example

Find the LCM of 6, 4, 3.

The multiples of 6 are 6, 12, 18, 24, 30, 36 . . .

The multiples of 4 are 4, 8, 12, 16, 20, 24, 28, 32 . . .

The multiples of 3 are 3, 6, 9, 12, 15, 18, 21, 24, 27, 30 . . .

The LCM of 6, 4, 3 is **12** ANS

Fill in this example:

Find the LCM of 7 and 5

The multiples of 7 are

The multiples of 5 are

The LCM of 7 and 5 is ANS

Now try these:

1. Find the LCM of 12, 15.
2. Find the LCM of 8, 12.
3. Find the LCM of 4, 6, 8.
4. Find the LCM of 2, 3, 5.

H Mixed Examples

Here are some mixed examples of this unit.
Look back over the previous work if you get stuck:

1. Express £21.148 to the nearest penny.
2. Express £8.512 to the nearest ten pence.

3. Express 8.3156 kg to the nearest gram. (Think about this one.)
4. Express 17.245 to 3 sig. fig.
5. Express 0.002 471 to 4 decimal places.
6. Arrange $\frac{3}{4}$, 0.7 and $\frac{3}{5}$ in order of size, smallest first.
7. Arrange 1.83, $1\frac{9}{11}$, 1.825 in order of size, biggest first.
8. Find $(40)^2$.
9. Find $\sqrt{49}$.
10. Find the HCF of 9, 12, 24.
11. Find the LCM of 5, 8, 10.
12. I have two pieces of wood one is 12 in long the other 9 in. I want to cut them both into several equal pieces with none left over.
    What is the biggest length I can cut?
    (A use for HCF.)
    Try to think of a useful example of LCM.

UNIT SIX **Test**

Set your work out neatly on lined paper. Put your name at the top.

1. Find the LCM of 6 and 9.
2. Find the HCF of 6 and 9.
3. Find $(1.5)^2$.
4. Find $\sqrt{81}$.
5. Arrange in order of size biggest first $2\frac{1}{3}$, 2.33, $2\frac{2}{5}$.
6. Arrange in order of size smallest first $\sqrt{36}$, 6.1, $\frac{11}{2}$.
7. £23 ÷ 11 to the nearest penny.
8. Find the value of $(1.2)^2$ to 1 decimal place.
9. Multiply 7.204 × 8 to 3 sig. fig.
10. Building blocks are in two lengths 3 cm and 4 cm. What is the smallest length I can make using either all 3 cm or all 4 cm blocks.

Do not forget to check your work.

A

b)	£104.70	l)	13 p
c)	£28.54	m)	46 p
d)	£5.27	n)	21 p
e)	£83.00	o)	13 p
f)	£16.79	p)	62 p
g)	£100.92	q)	£5.86
h)	£33.33	r)	29 p
i)	£0.03	s)	14 p
j)	£17.69	t)	62 p

B

1. 8.615 2. 17.12 3. 1.0
4. 8.25 5. 17.0

C

1. 136 2. 64 3. 0.012
4. 0.0870 5. 1000

D

1. 1.9, $1\frac{8}{9}$, 1.88 2. $\frac{5}{9}$, $\frac{5}{8}$, $\frac{6}{7}$
3. $3\frac{2}{11}$, 3.14, 3.139

E

$6\frac{1}{4}$		4	
625	NB $6\frac{1}{4} \times 100$	8	
10 000	NB 4 zeros	1000	NB Half as many zeros
50.41		30	
6400	NB 2 zeros	25	NB You have done 25^2

F

The factors of 36 are 1, 2, 3, 4, 6, 9, 12, 18, 36.
The factors of 60 are 1, 2, 3, 4, 5, 6, 10, 12, 15, 20, 30, 60.
The HCF of 36 and 60 is 12.

1. 18 2. 3 3. 1 4. 3

G

The multiples of 7 are 7, 14, 21, 28, 35, 42, 49.
The multiples of 5 are 5, 10, 15, 20, 25, 30, 35.
The LCM of 7 and 5 is 35.

1. 60 2. 24 3. 24 4. 30

H

1.	£21.15	2.	£8.50
3.	8.316 kg	4.	17.2
5.	0.0025	6.	$\frac{3}{5}$, 0.7, $\frac{3}{4}$
7.	1.83, 1.825, $1\frac{9}{11}$	8.	1600
9.	7	10.	3
11.	40	12.	3 in

UNIT SEVEN

Percentages

If you walk round a shopping centre looking in shop windows you will see signs like these:

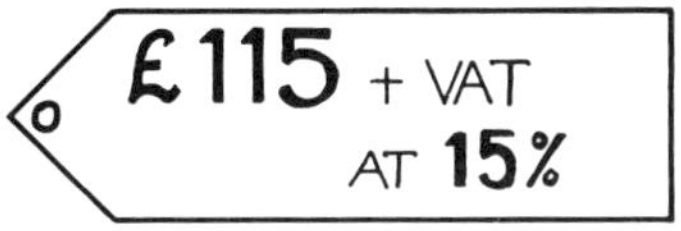

If you own a shop or business you will have to use percentages to work out VAT (Value Added Tax).

A Fractions to Percentages

To change a fraction to a percentage multiply by 100. 'Per cent' means 'for each hundred'.

Example

$$\frac{1}{2} = \frac{1}{2} \times \frac{100}{1}\% = \frac{100}{2}\% = 50\%$$

$$\frac{7}{8} = \frac{7}{\cancel{8}_2} \times \frac{\cancel{100}^{25}}{1}\% = \frac{175}{2}\% = 87\tfrac{1}{2}\%$$

You fill in these:

$$\frac{1}{4} = \frac{1}{4} \times \frac{100}{1}\% = \qquad\qquad \%\ \text{ANS}$$

$$\frac{3}{4} = \frac{3}{4} \times \qquad = \qquad\qquad \text{ANS}$$

$$\frac{1}{10} = \qquad\qquad = \qquad\qquad \text{ANS}$$

Now change these fractions to percentages:

1. $\frac{3}{8}$ 2. $\frac{1}{5}$ 3. $\frac{1}{3}$ 4. $\frac{3}{5}$ 5. $1\frac{1}{2}$

B Percentages to Fractions

To change a percentage to a fraction divide by 100. Cancel as much as possible.

Example

$$80\% = \frac{\cancel{80}^{4}}{\cancel{100}_5} = \frac{4}{5}$$

If the percentage contains a whole number and fraction turn it into an improper fraction and divide by 100.

$$12\tfrac{1}{2}\% = \frac{\cancel{25}^{1}}{2 \times \cancel{100}_4} \quad \textit{Then cancel.}$$

$$= \frac{1}{8}$$

Complete these:

$$70\% = \frac{70}{100} = \qquad\qquad \text{ANS}$$

$$33\tfrac{1}{3}\% = \frac{\qquad}{3 \times 100} = \qquad\qquad \text{ANS}$$

$$125\% = \frac{\qquad}{\qquad} = \qquad\qquad \text{ANS}$$

Now change these to fractions:

1. 95%	2. $37\frac{1}{2}\%$	3. 225%
4. 16%	5. $66\frac{2}{3}\%$	

C Change to a Percentage

Sometimes it is necessary to express one number as a percentage of another.

Example

Suppose you get 45 marks out of 50 in an exam and want to know your percentage.

Your mark is $\frac{45}{50}$ as a fraction.

Multiply by 100 to get a percentage.

$$\frac{45}{{}_{1}\cancel{50}} \times \frac{\overset{2}{\cancel{100}}}{1} = 90\%$$

(A good mark. This would get a Distinction!)

Example

A girl needs £80 for a holiday. She pays £25 deposit. What percentage deposit has she paid?

$$\frac{25}{{}_{4}\cancel{80}} \times \frac{\overset{5}{\cancel{100}}}{1} = \frac{125}{4} = 31\frac{1}{4}\%$$

She has paid $31\frac{1}{4}\%$ deposit ANS

What percentage is left to pay?

The whole amount is 100%.

$100 - 31\frac{1}{4} = 68\frac{3}{4}$

She still has $68\frac{3}{4}\%$ to pay ANS

Fill in these:

Express 25 g as a percentage of 1.5 kg

Change both to grams. 1.5 kg = g

Percentage is $\frac{25}{} \times \frac{100}{1} =$ ANS

Express 55 p as a percentage of £2.20

£2.20 = p

$\frac{}{} \times \frac{100}{1} =$ ANS

Now try these on paper:

1. A boy gets 8 out of 10 in a test. What is this as a percentage?
2. Express 500 g as a percentage of 2.5 kg.
3. Express 50 p as a percentage of £5.
4. A bottle contains 2 ℓ of lemonade. If a girl drinks 40 cℓ what percentage is left?
5. A house costs £30 000. A couple has £12 000 deposit. What percentage do they need to borrow to buy the house?

D Find a Percentage

You will sometimes need to find a percentage of a number or amount.

Example

To book a holiday you have to send 10% deposit. What is 10% of £150?

First write 10% as a fraction $= \frac{10}{100}$

Then you need $\frac{10}{100}$ **of £150**

$$= £\frac{\cancel{10}}{\cancel{100}} \times \frac{15\cancel{0}}{1} = £15$$

Set out your work like this:

Example

Find 15% of £120.

$$£\frac{\overset{3}{\cancel{15}}}{\underset{5}{\cancel{100}}} \times \frac{\overset{6}{\cancel{120}}}{1} = £18$$

Example

Find $12\frac{1}{2}$% of 500 kg.

$$\frac{12\frac{1}{2}}{100} \times \frac{500}{1}$$

$$= \frac{25}{2 \times 100} \times \frac{500}{1} = \frac{125}{2}$$

$$= 62\frac{1}{2}\text{ kg}$$

If you have to increase or decrease an amount by a percentage, work out the percentage and add or subtract from the original amount.

Example

Increase £40 by 10%.
Find 10% of £40.

$$\frac{10}{100} \times \frac{40}{1} = £4$$

Add it on.

$$£40 + £4 = £44$$

Example

Decrease 150 by 5%.

$$\frac{5}{{}_2100} \times \frac{150}{1} = \frac{15}{2} = 7\frac{1}{2}$$

Subtract. $150 - 7\frac{1}{2}$

$$= 142\frac{1}{2}$$

You try these:

Find 8% of 500.

$$\frac{8}{} \times \frac{}{1} = \qquad \text{ANS}$$

Find $37\frac{1}{2}$% of £64.

$$\frac{37\frac{1}{2}}{100} \times \frac{}{1} =$$

$$\frac{}{2 \times 100} \times \frac{}{} =$$

Decrease £600 by 20%.

$$\frac{20}{} \times \frac{}{}$$

$$= 600 - \qquad = \qquad \text{ANS}$$

Now try these on paper:

---

1. Find 12% of £500.
2. VAT is 15%. What is the VAT on an item costing £400 before VAT?
3. A sack of coal weighs 50 kg. If 5% is lost through a split what weight is left?
4. A house can be bought for 10% deposit. How much deposit is needed for a £45 000 house?
5. A teacher says you need 55% to pass an exam. If there are a possible 200 marks how many must you score to pass?
6. Increase £500 by 8%.
7. Decrease 50 kg by 25%.

E Simple Interest

Example

If a bank offers 8% interest per annum (year) and you invest (pay to bank) £200, then after one year you get 8% of £200 interest.

$$\textit{Interest} = \frac{8}{100} \times \frac{200}{1} = £16$$

If you spend the £16 but leave the £200 with the bank you will get £16 every year.
After 3 years you will have had 3 × £16 = £48 interest.
NB Simple Interest assumes you do not leave the interest in the bank to increase the £200.

Example

Find the Simple Interest on £400 invested for 5 years at 10%.

$$\text{Interest for 1 year} = \frac{10}{100} \times \frac{400}{1} = £40$$

$$\text{Interest for 5 years} = £40 \times 5$$

$$= £200$$

Fill in these examples:

Find the Simple Interest on £1000 invested for 4 years at 12%.

$$\text{Interest for 1 year} = \frac{12}{100} \times \frac{1000}{1} =$$

$$\text{Interest for 4 years} = 4 \times$$

$$= \qquad \text{ANS}$$

Find the Simple Interest on £200 invested for $2\frac{1}{2}$ years at $12\frac{1}{2}$%.

$$\text{Interest for 1 year} = \frac{12\frac{1}{2}}{100} \times \frac{\quad}{\quad}$$

$$= \frac{\quad}{2 \times 100} \times \frac{\quad}{\quad}$$

$$=$$

$$\text{Interest for } 2\tfrac{1}{2} \text{ years} = 2\tfrac{1}{2} \times$$

$$= \frac{5}{2} \times$$

$$= \qquad \text{ANS}$$

Now try these:

1. Find the Simple Interest on £500 for 3 years at 8%.
2. Find the Simple Interest on £250 for 2 years at 15%.
3. Find the Simple Interest on £5 for 3 years at 5%.
4. Find the Simple Interest on £25 for 10 years at 15%.
5. Find the Simple Interest on $250 for 2 years at 6%.

F Hire Purchase

If you wish to buy an expensive item you can sometimes pay a deposit and then a number of equal payments. This usually costs more than if you pay cash all at once.

Example

A furniture shop offers a wall unit for £250 cash or 24 monthly payments of £13.13 with no deposit.

a) *Calculate the total payment on HP.*

```
 £
  13.13×
     24
 ------
  52 52
 262 60
 ------
 315.12
```

The HP total is £315.12 ANS

b) *Find the extra cost.*

```
 £
 315.12−
 250.00
 ------
  65.12
```

The extra cost is £65.12 ANS

c) *Express the extra cost as a percentage of the cash price (to nearest whole number).*

$$\frac{65.12}{{}_5\cancel{250}} \times \frac{\overset{2}{\cancel{100}}}{1} = \frac{130.24}{5}$$

$$= 26.048$$

The interest charged is 26% ANS

Before you buy anything on HP always find out how much extra you will be paying!

Example

A bicycle is advertised at £63.00 or one third down and 52 weekly payments of £1.10.
What is the cost of the bicycle on HP?

Deposit $= \frac{1}{3} \times \frac{63}{1} = £21$ (63 and 3 cancelled: 21 and 1)

Payments = £

```
 1.10×  = £57.20
   52
 ----
 2 20
55 00
-----
57.20
```

Total cost £

```
21.00+
57.20
-----
78.20
```

The total cost is £78.20 ANS

Try this one:

A car costs £1000. It can be bought for 20% deposit and 12 monthly instalments of £70 each. Find the total cost on HP.

Deposit $= \frac{20}{100} \times \frac{1000}{1} =$

Payments = 70× =
12

Total cost +

The total HP cost is ANS

Now try this one:

A colour television costs £400. It can be bought on HP for one quarter down and 24 monthly payments of £15 each. Find the total HP cost and percentage interest charged.

Deposit $= \frac{1}{4} \times \quad =$

Payments = × =
24

Total HP cost +

The total HP cost is ANS

The interest charged is £
−
400

As a percentage of £400 this is

$\frac{\quad}{400} \times \frac{100}{1} =$

The percentage interest is % ANS

Now try these on paper:

1. A video recorder costs £500 cash or 20% down and 24 monthly payments of £20 each. Find the HP price and interest rate charged.
2. A dress shop offers £100 worth of goods for £20 deposit and 12 monthly payments of £6.67. Is this a good offer? Why?
3. "Shark Cars Ltd." offer a second hand car for £500 cash or 50% deposit and 12 monthly payments of £40 each. Find the percentage charged for HP. Is it a good offer?
4. Two shops have the same stereo unit for sale. Shop A asks for £20 deposit and 12 payments of £5.50
   Shop B offers £10 deposit and 12 payments of £6.50
   Which is cheaper and by how much?
5. You own a shop and have a washing machine for sale at £200. A customer asks for no deposit HP over 12 months. If you want to make 10% extra profit how much per month must you charge? (To nearest penny.)

G Mixed Examples

Now some mixed examples. Look back over the previous pages if you have forgotten anything:

1. Change $\frac{7}{8}$ to a percentage.
2. Change 115% to a fraction.
3. Express 81 marks out of 90 as a percentage.
4. Express £3.60 as a percentage of £5.
5. Find 20% of £400.
6. Increase £2.50 by 40%.
7. A factory makes 50 engines. If 8% are rejected how many are ready to sell?
8. Find the Simple Interest on £300 invested for 2 years at 12%.
9. Find the Simple Interest on £350 invested for 1 year 6 months at 5%.
10. A calculator costs £12.99 cash or £2.99 deposit and 12 monthly payments of 95 p. How much extra does HP cost?
11. A TV costs £300 or $33\frac{1}{3}$% deposit and 24 monthly payments of £9.85. How much extra is HP? Express this as a percentage of the cash price, to the nearest per cent.

When you have done these you are ready for Unit Seven Test.

UNIT SEVEN Test

Set your work out neatly on lined paper. Put your name at the top.

1. Change 85% to a fraction.
2. Change $1\frac{3}{4}$ to a percentage.
3. Express 250 g as a percentage of 4 kg.
4. Find 15% of £300.
5. Decrease £1000 by 8%.
6. A packing station rejects 12% of eggs as cracked.
   If 1200 eggs are delivered how many are likely to be cracked?
7. Find the Simple Interest on £5.50 invested for 5 years at 8%.
8. A milkman sells 120 bottles of milk in our road on Monday. On Friday he sells 140. What is the percentage increase?
9. A motorbike costs £499 cash or £125 deposit and 12 monthly payments of £38.50. How much extra does HP cost?
10. A television costs £350 cash or 20% down and 36 monthly payments of £9.99. How much extra is the HP as a percentage? (To the nearest per cent.)

Do not forget to check your work.

A $\frac{1}{4} = 25\%$, $\frac{3}{4} = 75\%$, $\frac{1}{10} = 10\%$

1. $37\frac{1}{2}\%$ 2. 20% 3. $33\frac{1}{3}\%$
4. 60% 5. 150%

B $70\% = \frac{7}{10}$, $33\frac{1}{3}\% = \frac{1}{3}$, $125\% = 1\frac{1}{4}$

1. $\frac{19}{20}$ 2. $\frac{3}{8}$ 3. $2\frac{1}{4}$
4. $\frac{4}{25}$ 5. $\frac{2}{3}$

C 1.5 kg = 1500 g

$$\frac{{}^{5}\not{25}}{{}_{3}\not{1500}} \times \frac{\not{100}}{1} = \frac{5}{3} = 1\frac{2}{3}\%$$

£2.20 = 220 p

$$\frac{{}^{5}\not{55}}{\not{220}_{\not{20}}} \times \frac{{}^{5}\not{100}}{1} = 25\%$$

1. 80% 2. 20% 3. 10%
4. 80% 5. 60%

D

$$\frac{8}{100} \times \frac{500}{1} = 40$$

$$\frac{37\frac{1}{2}}{100} \times \frac{64}{1} = \frac{{}^{3}\not{75}}{\not{2} \times \not{100}_{4}} \times \frac{{}^{\not{16}\,8}\not{64}}{1}$$

$$= £24$$

$$\frac{20}{\not{100}} \times 6\not{00} = £120$$

$$\begin{array}{r} 600- \\ 120 \\ \hline £480 \end{array}$$

1. £60 2. £60 3. 47.5 kg
4. £4500 5. 110 6. £540
7. 37.5 kg

E

$$\frac{12}{1\not{00}} \times \frac{100\not{0}}{1} = 120$$

$$4 \times 120 = £480$$

$$\frac{12\frac{1}{2}}{100} \times \frac{200}{1} = \frac{25}{\not{2} \times \not{100}} \times \frac{\not{200}}{1}$$

$$= £25$$

$$2\frac{1}{2} \times 25 = \frac{5}{2} \times \frac{25}{1} = \frac{125}{2}$$

$$= 62\frac{1}{2}$$

$$= £62.50$$

1. £120 2. £75 3. 75 p
4. £37.50 5. $30

F

$$\frac{20}{1\not{00}} \times \frac{100\not{0}}{1} = £200$$

$$\begin{array}{r} 70\times \\ 12 \\ \hline 840 \end{array} = £840$$

Total cost = **£1040 ANS**

$\frac{1}{4} \times 400 = £100$

$£15 \times 24 = £360$

Total HP cost = **£460 ANS**

Interest = £460 – £400 = £60

$$\frac{{}^{15}\not{60}}{\not{400}} \times \frac{\not{100}}{1} = 15\%$$

1. £580, 16%
2. Yes, HP is only 4 p more.
3. HP price = £730. 46%. An expensive offer.
4. A costs £86, B costs £88. A is cheaper by £2.
5. £18.33

G

1. $87\frac{1}{2}\%$
2. $1\frac{3}{20}$
3. 90%
4. 72%
5. £80
6. £3.50
7. 46
8. £72
9. £26.25
10. £1.40
11. £36.40, 12%

UNIT EIGHT

Averages and Revision

A Averages

To find the average of a group of numbers add them up and divide by the number of items.

Example

Find the average of 7, 9, 11, 12, 15.

7 + 9 + 11 + 12 + 15 = 54

There are five items.

Average $\frac{54}{5} = 10\frac{4}{5}$

Other names for average that you may meet are mean, and arithmetic mean.

Fill in this example:

Find the average length of 5.25 cm, 18.9 cm, 0.89 m and 6.54 m. (Answer to nearest millimetre).

Write them all in centimetres and add.

5.25

There are four items. 4)

ANS =

Now try these on paper:

1. Find the average weight of 5.2 kg, 6.9 kg and 2500 g to nearest gram.
2. Find the average cost of ten items at the tuck shop, whose prices are 5 p, 8 p, 8 p, 9 p, 10 p, 11 p, 12 p, 13 p, 14 p, 14 p. (Answer to one decimal place.)
3. A man's lunches for a week cost £1.10, £1.15, 85 p, 75 p, £1.55. Find the average cost.
4. The ages of six boys are 15 yr 2 m, 16 yr 5 m, 15 yr 8 m, 17 yr 0 m, 17 yr 5 m, 16 yr 4 m. Find their average age.
5. The temperatures during a week's holiday were:

	S	M	T	W	Th	F	S
deg C	22	24	25	21	20	18	22

   Find the average temperature correct to one decimal place.
6. A dealer sells four cars for £1250, £3200, £4500, £2850. Find the average price.
7. If I buy two books costing £1.99 each, and three books costing £2.99 each, find the average cost. [NB Add up all five prices].

B Revision

You have been working on the course for several weeks. If you have understood each section you should be able to remember most of the methods. To let you see how you are progressing here are some exam-type questions.

You may do your working on the page or a piece of paper which you hand in separately.

Things to remember:

1. Write the answers clearly.
2. State the units if necessary.
3. If you alter an answer cross it out and write it again clearly.

4. Don't give alternatives as the examiner will always mark the wrong one.
5. Spend a minute or two on each question. If you can't do it leave it and go on to the next one.
6. When you get to the end, go through again, check each answer (by another method if possible). If you think a question is wrong do a third check before altering it. Have another try at any questions you left out.

Revision Self-Test No 1

Time yourself but don't rush.

Check your answers:

Start Time ________

Finish Time ________

Time Taken ________

1. $16 + 18 + 114 + 6 =$
2. $941 - 75 =$
3. $115 \times 19 =$
4. $625 \div 25 =$
5. $1\frac{1}{2} + 5\frac{3}{5} =$
6. $2\frac{3}{4} - 1\frac{5}{8} =$
7. $5\frac{1}{2} \times 3\frac{1}{4} =$
8. $6\frac{1}{2} \div 4\frac{1}{2} =$
9. £2.24 × 8 =
10. £89.04 ÷ 8 =
11. $6.82 + 5.9 + 15.042 =$
12. $17 - 9.99 =$
13. $2.62 \times 1.4 =$
14. $2.98 \div 100 =$
15. $10 \div 0.2 =$
16. $1\frac{3}{4} + 10\frac{2}{9} =$
17. £28 − £5.62 =
18. $3.14 \times 11 =$
19. $17 \div 4 =$
20. Find the cost of 124 stamps at 15 p each.
21. $26.4 \times 100 =$
22. $(0.4)^2 =$
23. What percentage of 500 is 25?
24. Find 15% of 200.
25. Find the cost of $3\frac{1}{2}$ kg of butter at £1.84 per kilogram.
26. Change £27 to francs if £1 = 10.90 francs.
27. 42 ℓ ÷ 5 (Answer in centilitres.) =
28. Arrange in order of size, smallest first $1\frac{2}{3}$, 1.65, $1\frac{7}{9}$.
29. Find the Simple Interest on £400 invested for 3 years at 8%.
30. A bicycle is advertised at £90 or one third down and 52 weekly payments of £1.50. Find the HP cost of the bicycle.

Mark your work from the answers.

Look at each question you got wrong and see if you can see what you did wrong.

Spend some time revising any topic you are not sure about.

Now try this test. This time, allow yourself only 1 hour.

Revision Self-Test No 2

Start time ________

Finish Time ________

1. $17 + 21 + 12 + 144 =$
2. $63 \times 13 =$
3. $2\frac{3}{4} - 1\frac{5}{8} =$
4. $6\frac{1}{2} \div 1\frac{1}{2} =$
5. £6.72 × 7 =
6. £9.04 ÷ 8 =

7. 3 m 15 cm × 10 (Answer in metres and centimetres.) =
8. 17 ℓ ÷ 5 (Answer in millilitres.) =
9. Find 12% of £5.
10. Find the average of 63 p, 11 p, 19 p, 25 p, 82 p.
11. How many 5 p pieces do I need to make £5.25?
12. Convert 18.2 kg to pounds if 1 kg = 2.2 lb.
13. Express 60% as a fraction.
14. A person earns £86 per week. What is their annual salary to the nearest £100? (1 year = 52 weeks.)
15. What is the total capacity of 200 yoghurt pots each holding 150 mℓ? (Answer in litres.)
16. Calculate the Simple Interest on £850 invested for 4 years at 5%.
17. Find $(2.8)^2$.
18. Write in order of size, largest first:
$\frac{7}{3}$, $2\frac{2}{9}$, 2.3
19. A woman earns £450 per month. She pays 8% to a pension fund. How much is left?
20. A clerk works 32 hours a week for four weeks. He earns £1.70 an hour. What is his total wage for the four weeks?

Mark your work and again revise where necessary.

Unit Eight test will contain a selection of the work you have done so far. Attempt it when you are ready.

Calculators

From Unit Fourteen onwards you can use a calculator. You will be putting yourself at a disadvantage if you do not have one.

If you have not got one already try to get one for Christmas.

For Arithmetic a simple calculator is all that is required. Make sure it has square roots. If you hope to go on to more advanced work then think about getting a scientific calculator (not a programmable one). It will not help you to have one that is too complicated for you to use easily.

UNIT EIGHT **Test**

Set your work out neatly on lined paper. Put your name at the top.

1. $\frac{341 + 259}{150} =$
2. 0.77 + 70.7 + 7.7 =
3. $4\frac{1}{2} \div \frac{3}{4} =$
4. Work out the Simple Interest on £750 for $2\frac{1}{2}$ years at 6%.
5. If £1 = $1.58 then change £350 to dollars.
6. Find the average of £2.45, £5.50, £6.48, £5.45 and £1.32.
7. A camera is sold for £105 cash or one third down and 24 monthly payments of £4.65. Find the extra cost of using HP.
8. A family uses three pints of milk each weekday and four pints on Saturday and Sunday. How many gallons do they use in a year? (1 year = 52 weeks.)
9. A factory produces items weighing 125 g each. 144 items go in a box. The empty box weighs 200 g. What does the full box weigh? (Answer in kilograms.)
10. A car costs £5000 new. After 2 years it is sold for £3500. Find the loss as a percentage of the cost price.

Do not forget to check your work.

A

$$\begin{array}{r} 5.25 \\ 18.90 \\ 89.00 \\ 654.00 + \\ \hline 767.15 \\ \hline \end{array}$$

$$\begin{array}{r} 191.7875 \\ 4\overline{)767.1500} \end{array}$$

ANS = **191.8 cm** to nearest millimetre.

1. 4867 g or 4.867 kg	**2.** 10.4 p
3. £1.08	**4.** 16 yr 4 m
5. 21.7 °C	**6.** £2950
7. £2.59	

Revision Self-Test 1

1. 154	**2.** 866
3. 2185	**4.** 25
5. $7\frac{1}{10}$	**6.** $1\frac{1}{8}$
7. $17\frac{7}{8}$	**8.** $1\frac{4}{9}$
9. £17.92	**10.** £11.13
11. 27.762	**12.** 7.01
13. 3.668	**14.** 0.0298
15. 50	**16.** $11\frac{35}{36}$
17. £22.38	**18.** 34.54
19. $4\frac{1}{4}$	**20.** £18.60
21. 2640	**22.** 0.16
23. 5%	**24.** 30
25. £6.44	**26.** 294.3 francs
27. 840 cℓ	**28.** 1.65, $1\frac{2}{3}$, $1\frac{7}{9}$
29. £96	**30.** £108

Revision Self-Test 2

1. 194	**2.** 819
3. $1\frac{1}{8}$	**4.** $4\frac{1}{3}$
5. £47.04	**6.** £1.13
7. 31 m 50 cm	**8.** 3400 mℓ
9. 60 p	**10.** 40 p
11. 105	**12.** 40.04 lb
13. $\frac{3}{5}$	**14.** £4500
15. 30 ℓ	**16.** £170
17. 7.84	**18.** $\frac{7}{3}$, 2.3, $2\frac{2}{9}$
19. £414	**20.** £217.60

UNIT NINE

Ratio and Proportion

A Costs

Given the cost of a number of items find the cost of a different number. This can be useful when comparing prices of goods packed in different amounts.

Example

If eight cakes cost £2, find the cost of five cakes.

Write down the information.

8 cakes cost **£2 = 200 p**

Find the cost of one cake.

No need to simplify yet.

1 cake costs $\frac{200}{8}$ (= **25 p**)

Five cakes cost five times as much.

Simplify.

5 cakes cost $\frac{\cancel{200}^{25}}{{}_{1}\cancel{8}} \times \frac{5}{1}$

= **125 p**

= **£1.25**

Answer in a sentence.

Five cakes cost £1.25 ANS

Example

If I pay £4.28 for 6 kg of washing powder, what is the comparable price for a $1\frac{1}{2}$ kg packet?

Write down the information (price at the end).

6 kg cost **£4.28 = 428 p**

Find the cost of 1 kg.

1 kg costs $\frac{428}{6}$

Now find $1\frac{1}{2}$ times as much.

$1\frac{1}{2}$ kg cost $\frac{428}{6} \times 1\frac{1}{2}$

$= \frac{\cancel{428}^{214}}{{}_{2}\cancel{6}} \times \frac{\cancel{3}^{1}}{\cancel{2}}$

$= \frac{214}{2} = 107$ p

$1\frac{1}{2}$ kg would cost £1.07 ANS

Fill in the examples:
If 5 kg of apples cost £2.20 what will 8 kg cost?

5 kg cost p

1 kg costs ———

8 kg costs ——— $\times \frac{8}{1}$ =

8 kg of apples cost ANS

If 150 parcels weigh 9 kg, what will 250 similar parcels weigh?

150 parcels weigh **9 kg**

1 parcel weighs $\frac{\quad}{\mathbf{150}}$ **kg** — *Don't work it out! It cancels later.*

250 parcels weigh $\frac{\quad}{\quad} \times \frac{\mathbf{250}}{\mathbf{1}}$

$=$

250 parcels weigh kg ANS

If you understand the method try these on paper. Write out the method as before. This will enable you to learn the method easily:

1. If 9 books cost 81 p what will seven cost?
2. If eggs cost 90 p a dozen what will 18 cost?
3. If 7 lb of potatoes cost 84 p what will 10 lb cost?
4. If I can fill 26 glasses from four bottles of wine, how many can I fill from six bottles?
5. If five people having a meal pay a total of £16.50, what will three people have to pay for a similar meal?

B Ratio

Ratio is written like this :
You may see it on maps. 1 : 25 000 means the map is drawn $\frac{1}{25\,000}$ of real size.

Example

A calculator costs £5 and Andrew and Brian decide to buy it between them. Andrew has £2, Brian has £3. How is ownership of the calculator divided?
They own the calculator in the ratio £2 : £3

= 2 : 3

Andrew owns **2 parts** and Brian 3 ANS

Example

Express 75 p to £1.20 as a ratio in lowest terms.

75 p : £1.20	*Change to same units.*
75 : 120	*Leave out units.*
15 : 24	*Divide both by 5 (like cancelling).*
5 : 8 ANS	*Divide by 3.*
	Nothing else divides into both.

Try these:

1. £5 : £1.25
 = 500 :
 = :

2. 150 cm : 2.5 m
 = 150 :
 = :

3. 40 cℓ: 1.2ℓ
 = :
 = :

4. £2200 : £1800
 = :
 = :

C Sharing in a given ratio

Example

In the previous example Andrew and Brian own a calculator in the ratio 2 : 3.
If there are 30 weeks of the school year left how long should each use the calculator?

Divide 30 into 5 parts. (2 + 3 = 5)

30 ÷ 5 = 6

Andrew keeps it for 2 × 6 weeks
= **12 weeks** (He has 2 shares.)

Brian keeps it for 3 × 6 weeks
= **18 weeks** (He has 3 shares.)

30 (Check total is 30)

Example

Divide £504 in the ratio 2 : 7.

There are 9 parts.

$$\begin{array}{r} 56 \\ 9\overline{)504} \end{array}$$

Each part is £56.

$$\begin{array}{r} 56\times \\ 2 \\ \hline 112 \\ \hline \end{array} \qquad \begin{array}{r} 56\times \\ 7 \\ \hline 392 \\ \hline \end{array} \qquad \text{Check} \quad \begin{array}{r} 112+ \\ 392 \\ \hline 504 \\ \hline \end{array}$$

ANS = £112 : £392

Fill in this one: *Divide £120 between A, B, C in the ratio 1 : 2 : 3.*

$\overline{)120}$ **1 + 2 + 3 =**   parts

Add to check

$$\begin{array}{r} \times \\ 1 \\ \hline \\ \hline \end{array} \qquad \begin{array}{r} \times \\ 2 \\ \hline \\ \hline \end{array} \qquad \begin{array}{r} \times \\ 3 \\ \hline \\ \hline \end{array}$$

ANS = £   : £   : £

Now try these on paper:

1. Divide 80 p in the ratio 3 : 5.
2. Divide £150 in the ratio 1 : 3 : 6.
3. Share 20 apples between Ann and Barbara in the ratio 2 : 3.
4. A mixture consists of flour, sugar and margarine in the ratio 2 : 1 : 1.
   How much of each is in 1 kg of the mixture? (Answer in grams.)
5. A line is 3 m long. Divide it in the ratio 1 : 2 : 3. (Answer in centimetres.)
6. Sand and cement are mixed in the ratio 2 : 1. How much sand is in 200 kg of mixture? (Answer in kilograms.)

D More Examples

Some money is divided in the ratio 3 : 4.
The second person gets £100. How much money was divided?

4 parts = **£100**

1 part = $\dfrac{100}{4}$ = **£25**

3 + 4 = 7

7 parts = **25 × 7 = £175**

£175 was divided ANS

Two people invest £1000 and £1500 in a factory. The profit is £50. How much should each receive?

Ratio of shares = **£1000 : £1500**
= **1000 : 1500**
= **2 : 3**

Share £50 in the ratio 2 : 3

$$\begin{array}{r} 10 \\ 5\overline{)50} \end{array}$$

$$\begin{array}{r} 10\times \\ 2 \\ \hline 20 \\ \hline \end{array} \qquad \begin{array}{r} 10\times \\ 3 \\ \hline 30 \\ \hline \end{array} \qquad \text{check} \quad \begin{array}{r} 20+ \\ 30 \\ \hline 50 \\ \hline \end{array}$$

They receive £20 and £30 ANS

You can do these:

1. *A dried fruit mixture consists of currants and sultanas in the ratio 3 : 2.*
   What weight of sultanas is needed with 105 g currants?

   3 parts = **105 g**

   1 part =

   Sultanas = 2 parts = **2×**

   g of sultanas are needed ANS

2. *Two people invest £1200 and £1500 in a business. The first gets £60 profit. How much does the second get?*

Ratio of shares = **£1200 : £1500**

= :

= :

parts = **£60**

1 part =

The second has parts = ×

=

The second gets £ ANS

Now try these:

3. Share 3000 ℓ of fuel between three aircraft in the ratio 2 : 1 : 3.
4. A, B and C invest £1000, £1000 and £2000 in a business. If B gets £50 profit what was the total profit?
5. Some money is divided in the ratio 5 : 3. If the second person gets £1.50, how much does the first get?

E Mixtures

Example

If a saleswoman mixes three sacks of cheap lawn seed at 20 p per kilogram with two sacks of expensive seed at 50 p per kilogram, how much has the mixture cost per kilogram?

Mixture is in ratio **3 : 2**

Every 5 kg cost **3 × 20 p + 2 × 50 p**

= 60 p + 100 p

= 160 p

1 kg costs $\frac{160}{5}$ **= 34 p**

The mixture costs 34 p per kilogram ANS

Example

Peardrops cost 24 p, sherbet is 26 p and acid drops 30 p for a quarter pound.
A shopkeeper mixes them in a ratio 2 : 1 : 1.
What is the price of the mixture?

2 × 24 p = 48 p
1 × 26 p = 26 p
1 × 30 p = 30 p
104 p for 2 + 1 + 1 = 4 parts

$$4\overline{)104}\ \to 26$$

The cost is 26 p per quarter pound ANS

Fill in this example:

A grocer mixes three types of tea in the ratio 1 : 2 : 3. The teas cost £1.20, £1.10 and 90 p per kilogram respectively. Find the cost of 1 kg of mixture. (*Answer to nearest penny.*)

1.20×	1.10×	.90×	Add +
1	2	3	
___	___	___	___
___	___	___	___

1 + 2 + 3 = 6 6)‾‾‾‾‾

1 kg costs ANS

Now try these:

1. Find the cost of a mixture of seeds in the ratio 2 : 3 if the first costs 20 p per kilogram and the second 40 p per kilogram.
2. A shopkeeper has an assortment of toys to put in a brantub (i.e. he will charge the same price for them all). They originally cost 10 p, 15 p and 25 p in the ratio 5 : 2 : 1. What is the smallest price in whole pence that he can charge without making a loss?
3. The Girl Guides decide to sell squash at a fête. Orange squash is diluted with water in the ratio 1 : 5. The squash is 90 p per litre, the water free. Paper cups cost 1 p each and hold 200 mℓ . How much must they charge to make 2 p profit per drink?

F Mixed Examples

Here are some mixed examples for you to try. Look back at previous examples if you get stuck:

1. If five cakes cost 30 p what will eight cakes cost?
2. If it costs 45 p for a six-mile bus journey, what will a ten-mile journey cost?
3. Express 90 p to £1.20 as a ratio in the lowest terms.
4. Divide 50 sweets in the ratio 2 : 3.
5. Divide 20 gal of petrol between three cars in the ratio 2 : 3 : 3. (Answer in gallons and pints.)
6. In his will a man divided his money between his sons in the ratio 4 : 3 : 2. If the smallest amount a son received was £13 000 how much money did the man leave?
7. Two women start a business. A invests £25 000 and B £35 000. In the first year A gets £3000 profit. What does B get?
8. Three boys bought an old car between them in the ratio 5 : 4 : 3. The car cost £600. How much did each pay?
   After three months' hard work they sold it for £840. They shared the money in the same ratio. What did each receive?
9. Some money was shared in the ratio 3 : 4. The first person got £45. How much did the second get?
10. A basket of fruit contained apples, oranges and bananas in the ratio 5 : 4 : 2. If there were six bananas how many items were there altogether?

Mark your work from the answers and look back if you don't understand any question.
Then do Unit Nine Test without help from your notes.

UNIT NINE Test

Set your work out neatly on lined paper. Put your name at the top.

1. Express 80 p to £2.60 as a ratio in lowest terms.
2. If eight sweets cost £1.92 find the cost of seven sweets.
3. Divide £56 in the ratio 3 : 5.
4. A and B buy a business, A investing £3500 and B £1000. How much will each receive from £810 profit?
5. Divide £352 in the ratio 2 : 3 : 6.
6. The cost of an article is made up of labour and materials in the ratio 5 : 3. If an article costs £48 how much is for labour, and how much for materials?
7. Find the cost of 63 articles at seven for 5 p.
8. A pack of biscuits consists of plain, cream and chocolate biscuits in the ratio 3 : 2 : 1. If the pack contains $1\frac{1}{2}$ kg of biscuits, find the weight of each type. (Answer in grams.)
9. In question **8**, if plain biscuits cost 42 p per kg, creams cost 56 p per kg and chocolate biscuits cost 96 p per kg find the cost of the $1\frac{1}{2}$ kg pack. (Answer to the nearest penny.)
10. A man wins some money. He invests some, spends some and gives some to his friends in the ratio 5 : 4 : 1. If he gives £250 to his friends, how much does he invest?

Do not forget to check your work.

A 5 kg cost **£2.20 = 220 p**

1 kg costs $\frac{220}{5}$

8 kg cost $\frac{220^{44}}{5} \times \frac{8}{1} = 352$

8 kg of apples cost £3.52 ANS

150 parcels weigh **9 kg**

1 parcel weighs $\frac{9}{150}$ kg

250 parcels weigh $\frac{9^{3}}{{}_{3}150} \times \frac{250^{5}}{1}$

$= 15$ kg

250 parcels weigh 15 kg ANS

1. 63 p 2. £1.35 3. £1.20
4. 39 5. £9.90

B

1. 500 : 125 = 4 : 1
2. 150 : 250 = 3 : 5
3. 40 : 120 = 1 : 3
4. 2200 : 1800 = 11 : 9

C **1 + 2 + 3 = 6** parts

$6\overline{)120} = 20$

20 × 1	20 × 2	20 × 3	20 + 40 + 60
20	40	60	120

ANS = £20 : £40 : £60

1. 30 p : 50 p
2. £15 : £45 : £90
3. 8 : 12
4. 500 g : 250 g : 250 g
5. 50 cm : 100 cm : 150 cm
6. $133\frac{1}{3}$ kg

D

1. 3 parts = **105 g**

   1 part = $\frac{105}{3}$ = **35 g**

   Sultanas = **2 parts = 35 × 2**
   = 70 g

2. Ratio of shares = **£1200 : £1500**
   = 1200 : 1500
   = 4 : 5

   4 parts = **£60**

   1 part = **£15**

   The second has 5 parts = **£15 × 5**
   = £75

   The second gets £75 ANS

3. 1000 ℓ : 500 ℓ : 1500 ℓ
4. £200
5. £2.50

E

1.20×	**1.10×**	**.90×**	**1.20+**
1	**2**	**3**	**2.20**
1.20	2.20	2.70	**2.70**
			6.10

1 + 2 + 3 = 6 parts

$$6\overline{)610.00} = 101.66$$

6 kg cost **£6.10 = 610 p**

1 kg costs £1.02 ANS (to nearest penny)

1. 32 p per kilogram
2. 14 p
3. 6 p

F

1. 48 p
2. 75 p
3. 3 : 4
4. 20 : 30
5. 5 gal : 7 gal 4 pt : 7 gal 4 pt
6. £58 500
7. £4200
8. £250 : £200 : £150, £350 : £280 : £210
9. £60
10. 33 items

UNIT TEN

Area

A Area

A unit of area is a square centimetre.
It measures one centimetre in each direction.
The area of any shape is the number of square centimetres (cm^2) it takes to cover it.

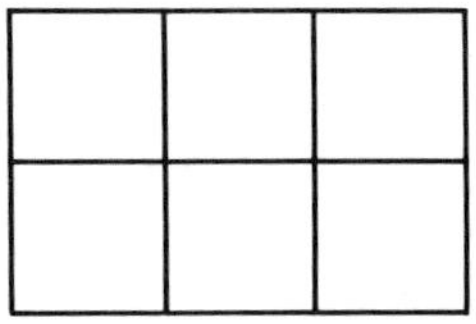

Area = 6 cm^2

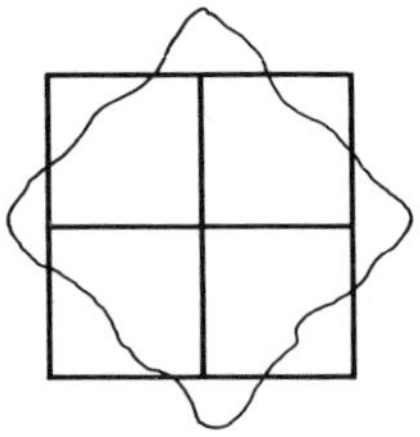

Area = approx 4 cm^2

Draw centimetre squares to find the area of these shapes.
(Draw the squares even if you know how to calculate the areas!)

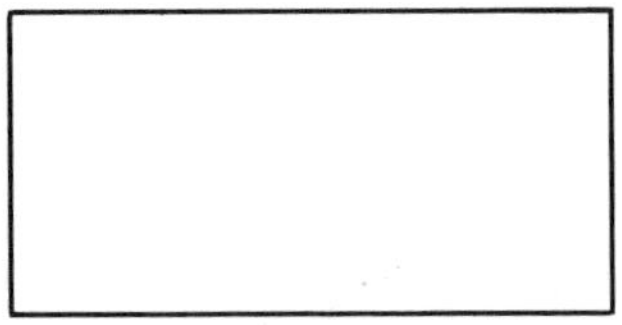

Area = cm^2

Area =

Area =

B Units

Any unit of length, e.g. in, m, km, can be used as a square unit for area i.e. in^2, m^2, km^2. In order to calculate an area you must have all the dimensions in the same units. If you are asked for an answer in cm^2 then work in cm. If you want an answer in m^2 then work in m.

Example

These show how to change units of areas.

Length		*Area*
10 mm = 1 cm	so	10×10 mm^2 = 1 cm^2
		100 mm^2 = 1 cm^2
100 cm = 1 m	so	100×100 cm^2 = 1 m^2
		10 000 cm^2 = 1 m^2
12 in = 1 ft	so	12×12 in^2 = 1 ft^2
		144 in^2 = 1 ft^2

Fill in the following:

1000 m = 1 km	so	$\times$ m^2 = 1 km^2
		m^2 = 1 km^2
ft = 1 yd	so	$\times$ ft^2 = 1 yd^2
		ft^2 = 1 yd^2
mm = 1 m	so	$\times$ mm^2 = 1 m^2
		mm^2 = 1 m^2

C Squares and Rectangles

The area of a square or a rectangle can be found by multiplying length $\times$ width.

> *Remember:*
> *Area of rectangle* $= l \times w$

Example

Find the area of a rectangular lawn width 10 m, length 15.5 m.

Area $= l \times w$

$= 15.5 \times 10$

$=$ **155 m^2**

Example

Find the area of a rectangle 8 in wide and 2 ft 6 in long. (Answer in square inches.)

Area $= l \times w$

$= 30 \times 8$ (2 ft 6 in = 30 in)

$=$ **240 in^2**

Fill in these examples:

1. *Find the area of a room 2.2 m wide and 4 m long.*

   Area $= l \times w$

   $=$

   $=$ m^2 ANS

2. *Find the area of a strip of paper 2 in wide and 3 yd long. (Answer in square inches.)*

   Area $= l \times w$

   $= 2 \times$

   $=$ in^2 ANS

Do these on paper:

3. Find the area of a square side 6 cm.
4. Find the area of a sheet of paper 21 cm wide and 30 cm long.
5. Find the area of a rectangle $2\frac{1}{2}$ in by $3\frac{1}{2}$ in.
6. Find the area of a rectangle 50 cm wide and 3 m long. (Answer in square metres.)

D Triangles

A triangle will always fit inside a rectangle and be half the area of the rectangle.

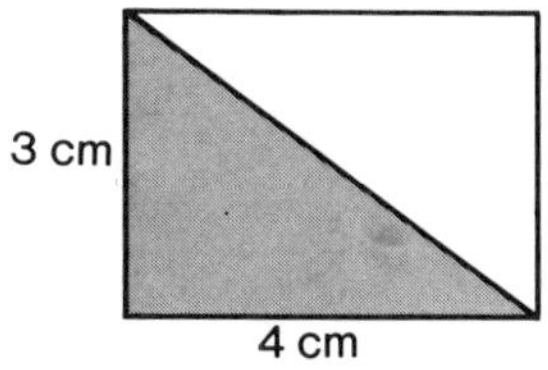

Area $= \frac{1}{2} \times 4 \times 3$

$= 6$ cm^2

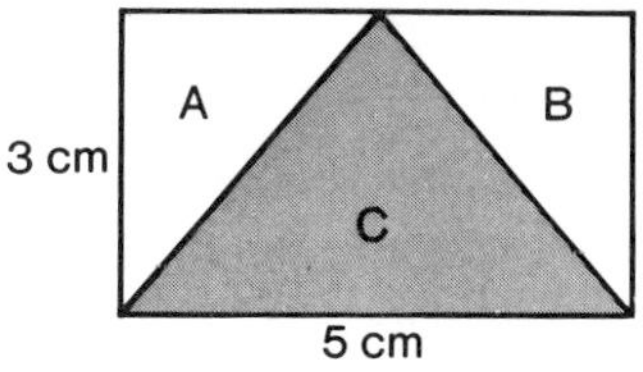

Area $= \frac{1}{2} \times 5 \times 3$

$= 7\frac{1}{2}$ cm^2

The two pieces A and B will exactly cover triangle C. (If you don't believe it copy them and cut them out!)

Measure these triangles and calculate their areas. Work in the units given. Measure to the nearest millimetre or the nearest $\frac{1}{8}$ in.

1.

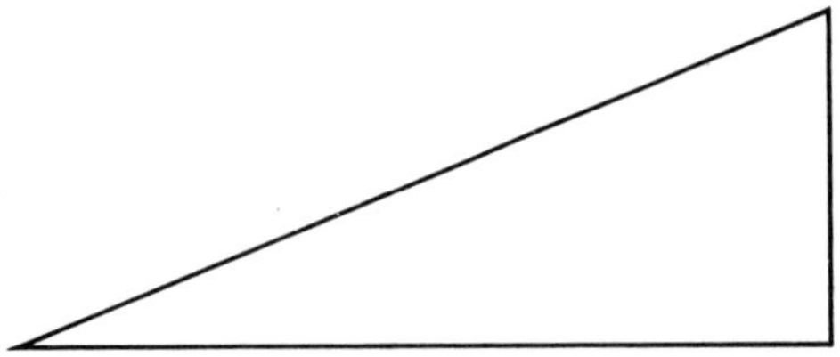

Area = cm^2

2.

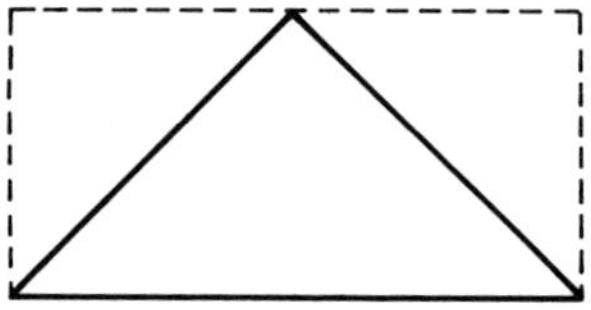

Area = in^2

3.

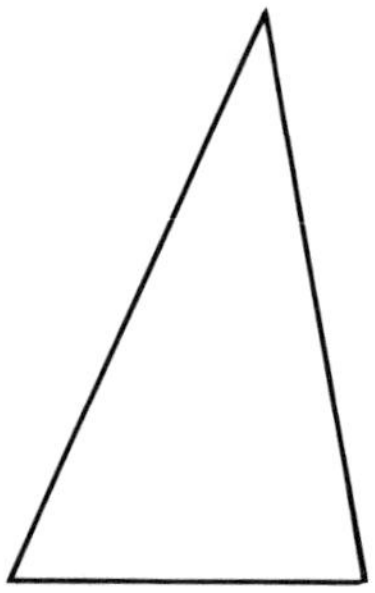

Area = cm^2

Area = in^2

4.

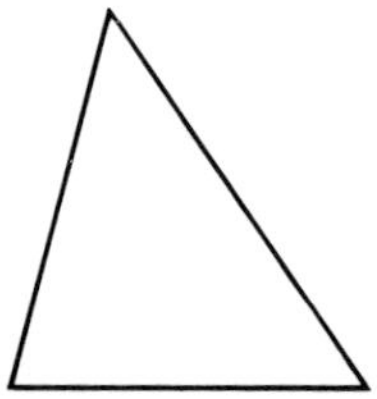

Area = mm^2

Area = cm^2

5.

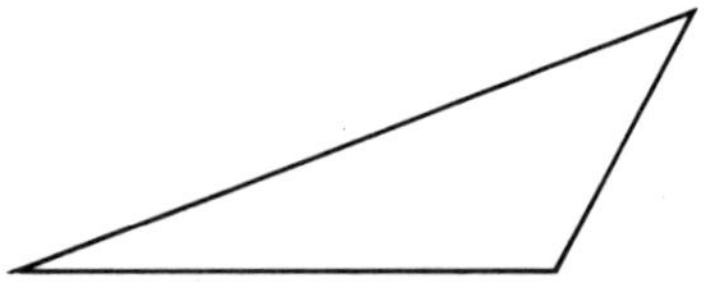

Area = in^2

Area = cm^2

6.

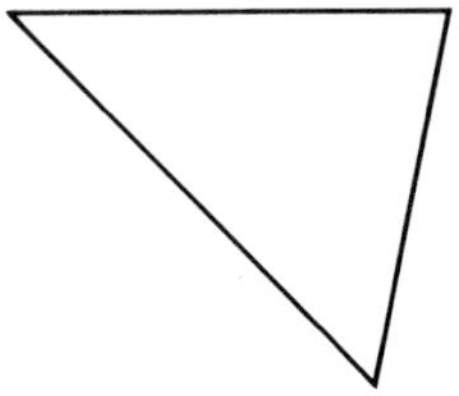

Area = cm^2

Area = mm^2

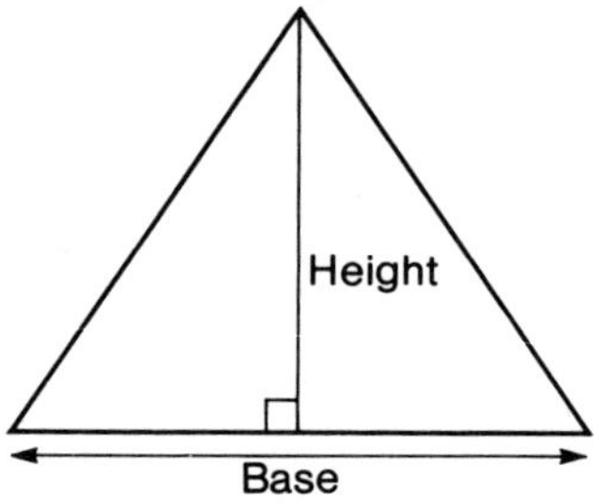

To find the area of a triangle you need to know the length of the base and height.

> *Remember:*
> *Area of triangle* $= \frac{1}{2} \times b \times h$

Sometimes you need to turn the diagram round.

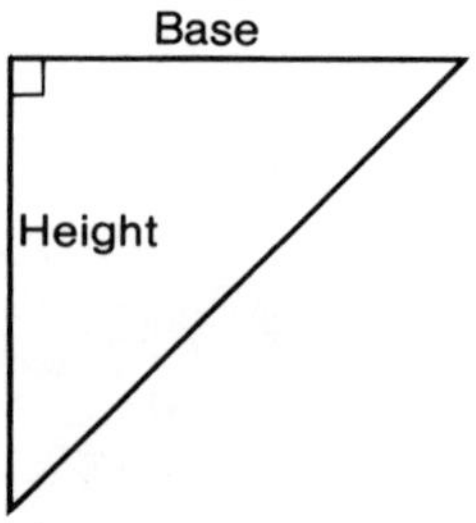

Example

(Fill in the gaps:)

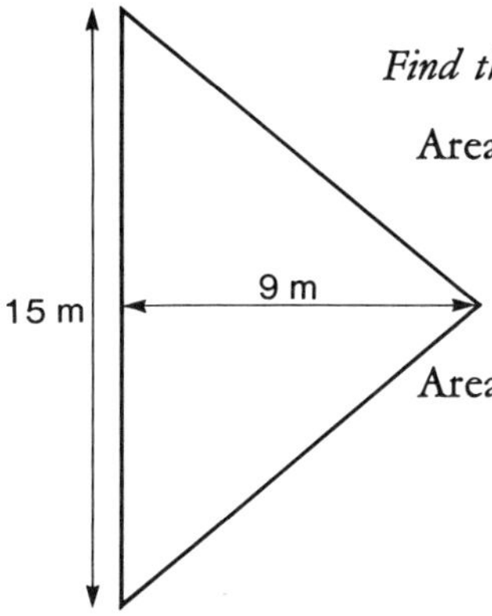

Find the area of this flower bed.

Area $= \frac{1}{2} \times b \times h$

$=$

$=$

Area $=$ $\qquad$ m^2

E The Circle

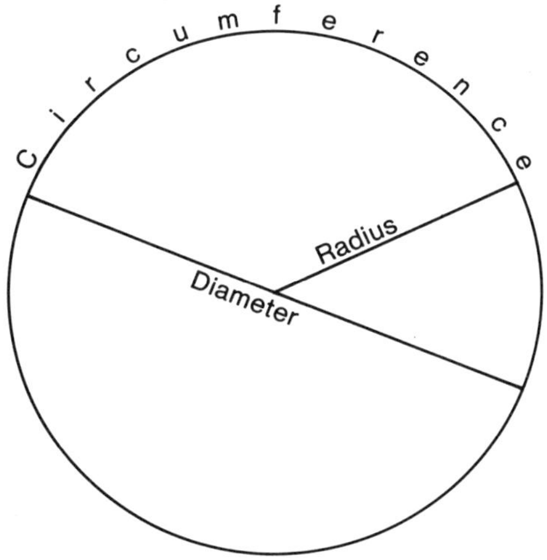

The circumference is the distance all round the circle. (It means the same as perimeter of a circle.)

The ratio of the circumference of a circle to its diameter is always the same. It is not an exact value but approximately 3.141 59 . . .

It has been given a special name Pi written π.
In arithmetic various numbers may be used for π.

$3\frac{1}{7}$, $\frac{22}{7}$, 3.14, 3.142

Usually it is best to use $\frac{22}{7}$ for a fractions question or where 7, 14, etc. occur and 3.14 otherwise.

If the problem tells you a value of π then use it. The area of a circle is πr^2.

> *Remember:*
> *Area of circle* $= \pi r^2 = \pi \times r \times r$

Write it down and learn it.

Example

Find the area of a circle radius 7 cm.

Area $= \pi \times r \times r$

$= \frac{22}{7} \times \frac{7}{1} \times \frac{7}{1} = 154$ cm^2

NB Use $\pi = \frac{22}{7}$ so the 7 cancels.
Don't forget the units — cm^2.

Example

Find the area of a circle radius 2.8 cm.

Area $= \pi \times r \times r$
$= 3.14 \times 2.8 \times 2.8$
$=$ **24.6176 cm^2** (by long multiplication).

Fill in these examples:

1. *Find the area of a circle diameter 7 cm.*

   Radius $=$ $\qquad$ (*use fractions*)

   Area $= \pi \times r \times r$

   $= \frac{22}{7} \times \quad \times$

   $=$ $\qquad$ cm^2 ANS

2. *Find the area of a circle radius 2.5 m.*

   Area $= \pi \times r \times r$

   $= 3.14 \times 2.5 \times 2.5$ $\qquad$ 3.14× $\qquad$ ×

   $=$ $\qquad$ m^2 $\qquad$ 2.5 $\qquad$ 2.5

Now work these out on paper:

3. Find the area of a circle with radius 35 cm $\left(\pi = \frac{22}{7}\right)$.

4. Find the area of a circle with radius 3 in ($\pi = 3.14$).

5. Find the area of a circle with diameter 49 m $\left(\pi = \frac{22}{7}\right)$.

6. Find the area of a circle with diameter 2 ft ($\pi = 3.14$).

F Working Backwards

Example

If the area of a rectangle is 12 cm² and the width is 2 cm find the length.

Area $= l \times w$ — *Write down formula.*

$12 = l \times 2$ — *Put in known numbers (check units are the same).*

$l = 12 \div 2$

$l = 6$ — *Find l.*

The length is **6 cm** ANS

Example

If the area of a square is 9 in² find the side.

Area $= l \times l$ — *Length and width are the same.*

$9 = l^2$ — *Find the square root.*

$l = 3$ The sides are **3 in** long ANS

Example

Find the diameter of a circle area 616 m² $\left(\pi = \frac{22}{7}\right)$

Area $= \pi \times r \times r$ — *Write down the formula.*

$$616 = \frac{22}{7} \times r \times r$$ *Put in numbers.*

$$r \times r = 616 \div \frac{22}{7}$$ *Divide both sides by $\frac{22}{7}$.*

$$= \overset{28}{\cancel{\overset{56}{\cancel{616}}}} \times \frac{7}{\cancel{22}_{\cancel{2}}}$$ *(Remember, how to divide fractions: if not look in Unit Three.)*

$$= 196$$

$$r = \sqrt{196} = 14$$ *Find the square root by guessing and squaring method.* (Harder examples use a calculator, see Unit Fourteen.)

The radius is **14 m** ANS

Now fill in these examples:

1. *If the area of a rectangle is 14.4 cm² and the length is 6 cm find the width.*

   Area $= l \times w$
   $14.4 =$
   $w = \quad \div \qquad)\overline{14.4}$
   $= \qquad$ cm ANS

2. *If the area of a triangle is 15 cm² and its base is 6 cm find the height.*

   Area $= \frac{1}{2} \times b \times h$
   $15 = \frac{1}{2} \times 6 \times h$
   $15 = 3 \times h$
   $h =$
   The height is $\qquad$ cm ANS

3. *If the area of a circle is 78.5 cm^2 find the radius* (π = *3.14*)

$$\text{Area} = \pi \times r \times r$$

$$78.5 = 3.14 \times r \times r$$

$$r^2 = 78.5 \div 3.14 \qquad 314\overline{)7850}$$

$$=$$

$$r =$$

The radius is cm ANS

Now try these examples on paper. Set them out neatly:

4. A triangular sheet of wood has a base 4 m and area 5.2 m^2. How high is it?
5. A rectangular carpet covers 14 m^2. If it is 4 m long how wide is it?
6. A gardener wants a circular flower bed with an area of 12 m^2. What must the radius measure? (Answer to nearest metre, $\pi = 3.14$.)
7. A cake recipe uses a 7 in diameter tin. Find the area of the cake. If I double the amount of mixture what size tin shall I use? $\left(\text{Answer to nearest inch, } \pi = \frac{22}{7}.\right)$
8. A tin of paint will cover 25 m^2. If the walls are 2.5 m high what width can be painted?

G Mixed Examples

Have you learnt all the formulae?
Fill them in then check back to make sure they are right.

Area of rectangle =
Area of triangle =
Area of circle =

Now try these mixed examples. Look to your notes if you get stuck:

1. Convert 1 km^2 to m^2.
2. Convert 1 yd^2 to ft^2.
3. Find the area of a rectangle $4\frac{1}{2}$ in by $2\frac{1}{2}$ in.
4. Find the area of a square of side 80 cm.
5. Find the area of a triangle base 15 cm and height 10 cm.
6. Find the area of a circle diameter 21 cm $\left(\pi = \frac{22}{7}\right)$.
7. Find the length of a rectangle width $2\frac{1}{2}$ m, area 75 m^2.
8. If nine tiles cover a square metre how many are needed to cover a floor 4.5 m by 12 m?
9. Find area. (Answer in square centimetres.)

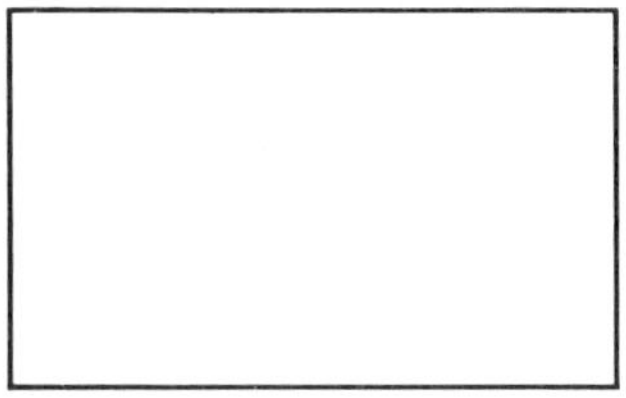

10. Find area. (Answer in square centimetres.)

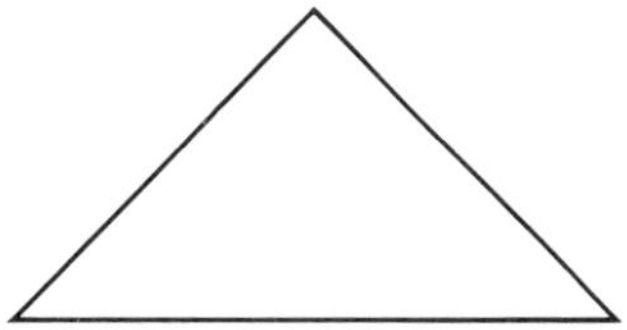

UNIT TEN **Test**

Set your work out neatly on lined paper. Put your name at the top.

Questions 1–3: Measure these shapes and find the areas in the units given:

1.

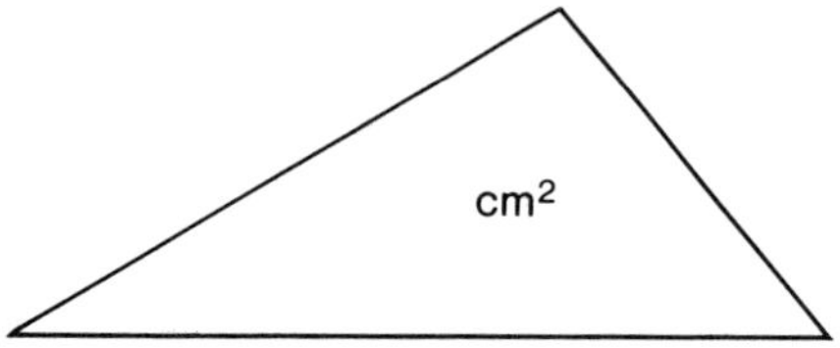

2.

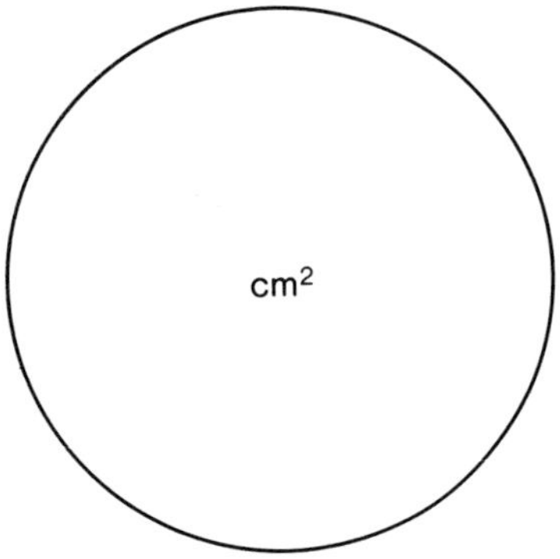

3.

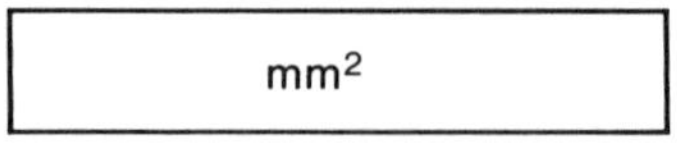

4. Find the width of a carpet area 36 m^2 and length 20 m. (Answer in centimetres.)
5. A circle covers 28.26 ft^2. Find its radius ($\pi = 3.14$).
6. Calculate the area of a circle of radius 7 cm $\left(\pi = \frac{22}{7}\right)$.
7. Calculate the area of a rectangle 12 cm wide and 16 cm long.
8. A triangle has an area 60 cm^2. The base is 20 cm. Find the height.
9. The diameter of a circular sports field is 210 m. Calculate the cost of returfing it at 30 p per square metre $\left(\pi = \frac{22}{7}\right)$.
10. A square has 40 ft sides. Find the area in square yards.

Do not forget to check your work.

A

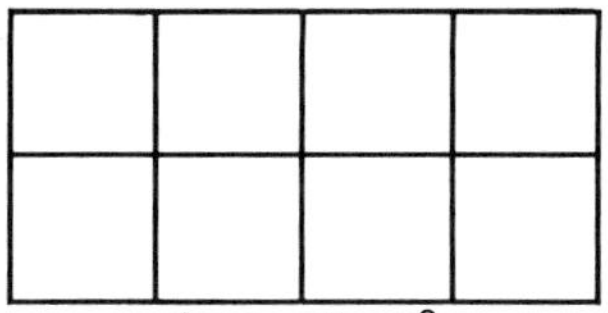

Area = 8 cm²

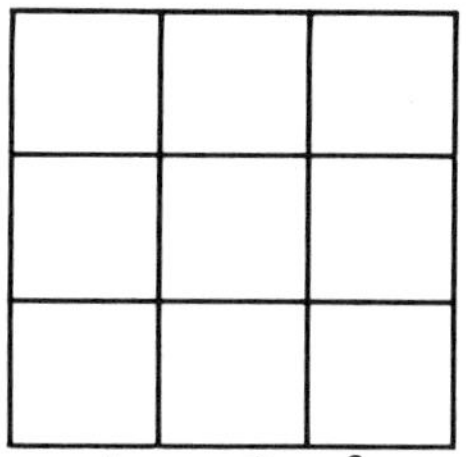

Area = 9 cm²

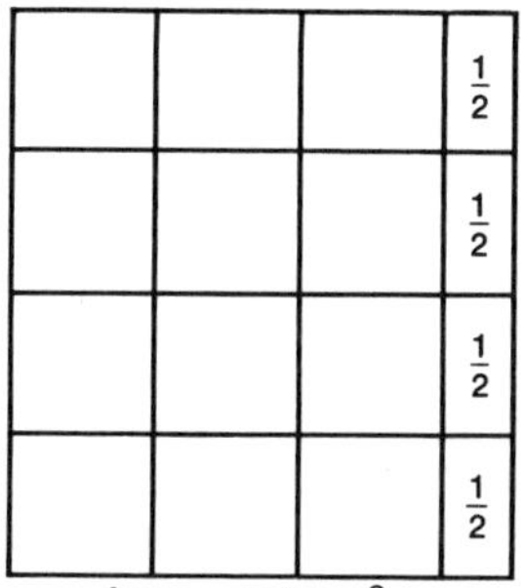

Area = 14 cm²

B

$$1000\text{ m} = 1\text{ km so } 1000 \times 1000\text{ m}^2 = 1\text{ km}^2$$
$$1\,000\,000\text{ m}^2 = 1\text{ km}^2$$
$$3\text{ ft} = 1\text{ yd so } 3 \times 3\text{ ft}^2 = 1\text{ yd}^2$$
$$9\text{ ft}^2 = 1\text{ yd}^2$$
$$1000\text{ mm} = 1\text{ m so } 1000 \times 1000\text{ mm}^2 = 1\text{ m}^2$$
$$1\,000\,000\text{ mm}^2 = 1\text{ m}^2$$

C

1. $\text{Area} = l \times w$
   $= 4 \times 2.2$
   $= \mathbf{8.8\ m^2}$

2. $\text{Area} = l \times w$
   $= 2 \times 108$
   $= \mathbf{216\ in^2}$

3. 36 cm² 4. 630 cm²

5. $\frac{5}{2} \times \frac{7}{2} = \frac{35}{4} = 8\frac{3}{4}\text{ in}^2$

6. $3 \times 0.5 = 1.5\text{ m}^2$

D

1. $\text{Area} = \frac{1}{2} \times 6 \times 2.5 = 7.5\text{ cm}^2$
2. $\text{Area} = \frac{1}{2} \times 1\frac{1}{2} \times \frac{3}{4} = \frac{9}{16}\text{ in}^2$
3. (i) $\text{Area} = \frac{1}{2} \times 2.3 \times 3.7 = 4.255\text{ cm}^2$
   (ii) $\text{Area} = \frac{1}{2} \times \frac{7}{8} \times 1\frac{1}{2} = \frac{21}{32}\text{ in}^2$
4. (i) $\text{Area} = \frac{1}{2} \times 24 \times 25 = 300\text{ mm}^2$
   (ii) $\text{Area} = \frac{1}{2} \times 2.4 \times 2.5 = 3\text{ cm}^2$
5. (i) $\text{Area} = \frac{1}{2} \times 1\frac{3}{8} \times \frac{3}{4} = \frac{33}{64}\text{ in}^2$
   (ii) $\text{Area} = \frac{1}{2} \times 3.5 \times 1.7 = 2.975\text{ cm}^2$
6. (i) $\text{Area} = \frac{1}{2} \times 2.8 \times 2.5 = 3.5\text{ cm}^2$
   (ii) $\text{Area} = \frac{1}{2} \times 28 \times 25 = 350\text{ mm}^2$

$\text{Area} = \frac{1}{2} \times b \times h$
$= \frac{1}{2} \times 15 \times 9$
$= \frac{1}{2} \times 135$
$= \mathbf{67.5\ m^2}$ or $\mathbf{67\frac{1}{2}\ m^2}$

E

1. radius $= 3\frac{1}{2}$ cm

area $= \frac{\cancel{22}^{11}}{\cancel{7}} \times \frac{\cancel{7}}{\cancel{2}} \times \frac{7}{2} = \frac{77}{2}$
$= 38\frac{1}{2}\text{ cm}^2$

2. Area $= 19.625\ m^2$

3. $3850\ cm^2$ 4. $28.26\ in^2$

5. $1886.5\ m^2$ or $1886\frac{1}{2}\ m^2$ 6. $3.14\ ft^2$

F 1. $14.4 = 6 \times w$ $\quad 6\overline{)14.4}$ = 2.4

$w = 14.4 \div 6$

$= \mathbf{2.4\ cm}$

2. 5 cm

3. $r^2 = 78.5 \div 3.14$

$r^2 = 25$

$r = \mathbf{5\ cm}$

$314\overline{)7850}$ = 25; 628; 1570; 1570; 0

4. 2.6 m 5. 3.5 m 6. 2 m

7. 10 in 8. 10 m

G Area of rectangle $= l \times w$ or lw

triangle $= \frac{1}{2} \times b \times h$ or $\frac{1}{2}bh$

circle $= \pi \times r \times r$ or πr^2

1. $1000\,000\ m^2$ 2. $9\ ft^2$

3. $11\frac{1}{4}\ in^2$ 4. $6400\ cm^2$

5. $75\ cm^2$ 6. $346\frac{1}{2}\ cm^2$

7. 30 m 8. 486 tiles

9. $10\ cm^2$ 10. $4\ cm^2$

UNIT ELEVEN

Perimeter and Volume

A Squares and Rectangles

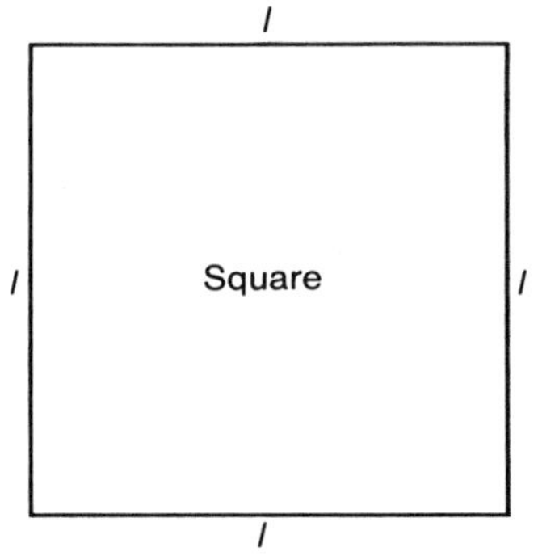

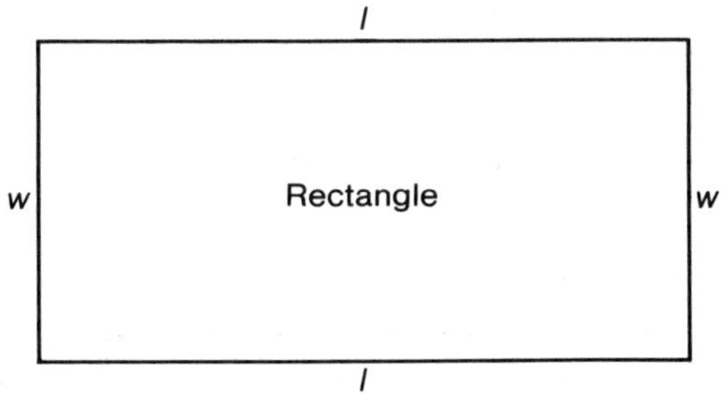

The perimeter is the distance all round the outside. If the sides of a square is l the perimeter $= 4 \times l$

The perimeter of a rectangle $= l + w + l + w$

Write down the perimeter of these shapes (don't forget the units — cm, in, etc.):

1.

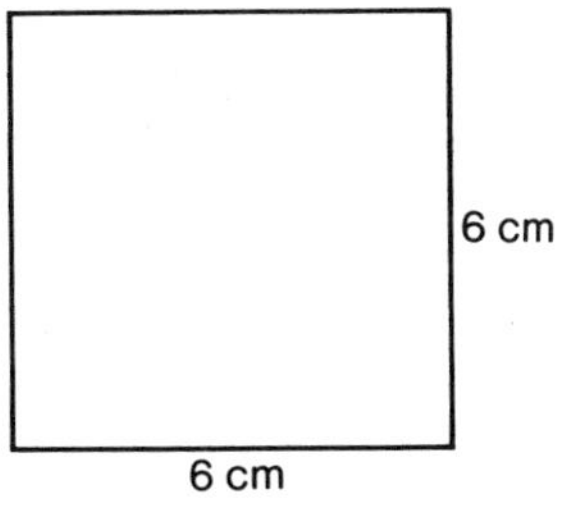

2.

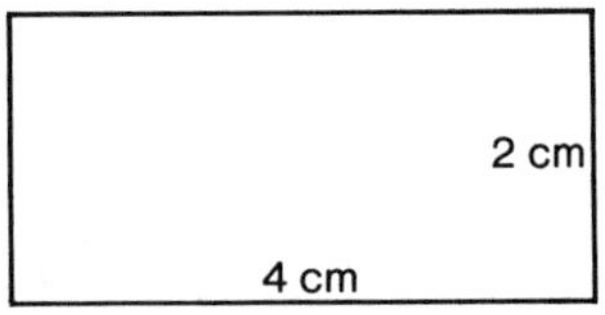

3.

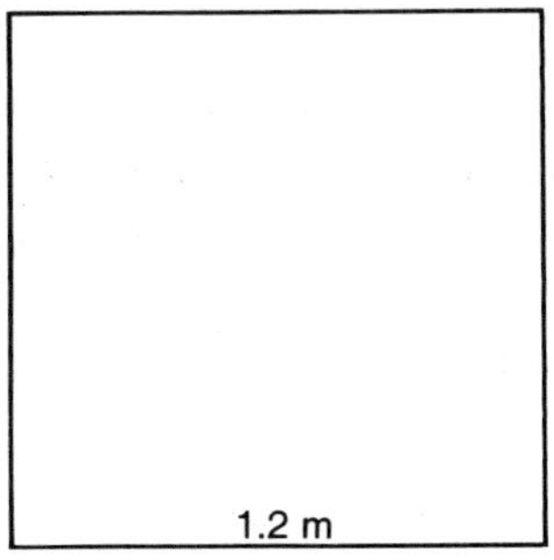

B Perimeter and Area

Example

If the perimeter of a square is 20 cm find its area.

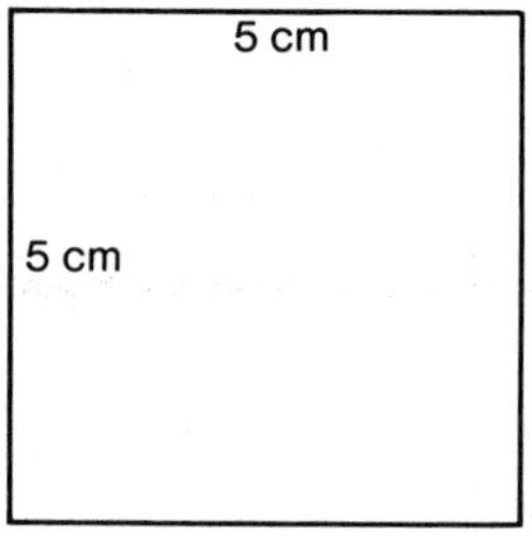

$P = 20$ cm

It helps to draw a sketch.

$l = 20 \div 4$	*Find the length of a side.*
$= 5$ cm	
$A = l \times w$	*Calculate the area.*
$= 5 \times 5$	
$= 25$ **cm²**	Don't forget the units.

Example

If the perimeter of a rectangle is 18 cm and the length is 5 cm find the area.

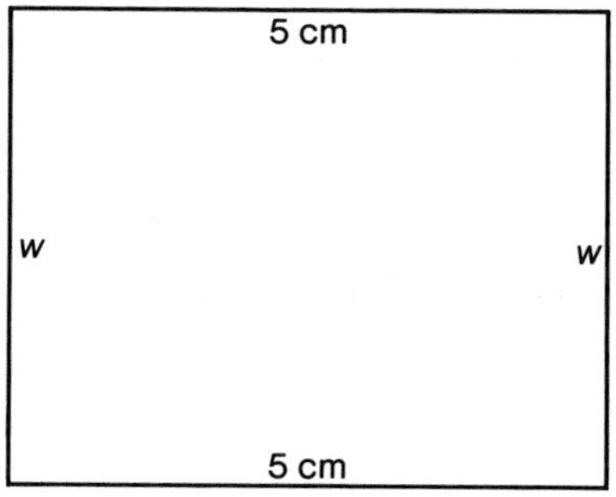

$P = l + w + l + w$	*Draw a sketch.*
$18 = 5 + w + 5 + w$	*We know* $l = 5$
$18 = 10 + w + w$	*We need to find the number for w.*

so

$w = 4$ cm	*Find the width.*
$A = l \times w$	*Find the area.*

$= 5 \times 4 = \mathbf{20\ cm^2}$

Fill in these examples:

1. *If the perimeter of a rectangle is 17 cm and the width is 2.5 cm find the area.*

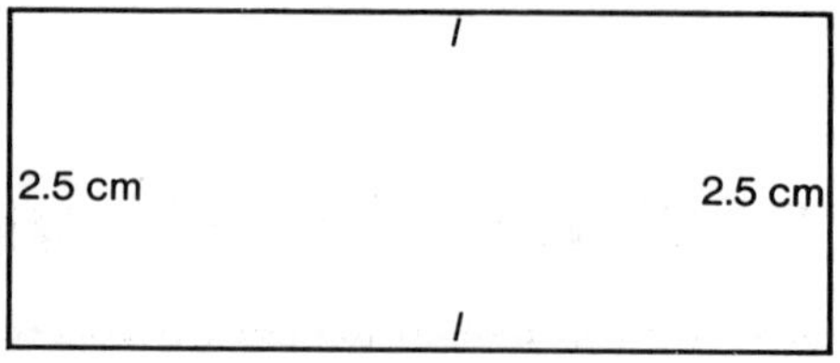

$P = 17 \text{ cm} = 2.5 + l + 2.5 + l$

$17 = l + l$

$l =$ cm

$A = l \times w$

$=$ ANS

2. *If the area of a square is 25 cm² find the perimeter.*

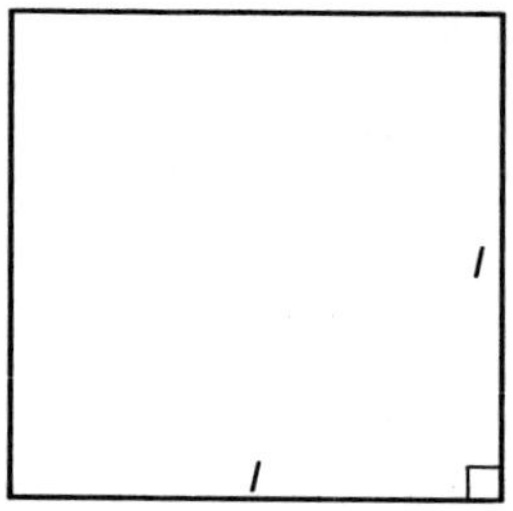

$A = l \times l = 25 \text{ cm}^2$

$l =$ cm

$P = 4l =$ ANS

Now do these on paper:

3. If the width of a carpet is 3 m and the area 15 m² find the perimeter.
4. If the perimeter of a square is 100 m find the area.
5. If the area of a rectangle is $15\frac{3}{4}$ cm² and the length $4\frac{1}{2}$ cm find the width and the perimeter.
6. If the perimeter of a square is 2.8 m find the area in square centimetres.

7. A square lawn has an area of 100 m^2. The gardener reduces the area by 19%. Find the side of the lawn before and after the alteration.
8. A floor measures 4.5 m by 3.2 m. It is to be covered with carpet. What area of carpet is needed? How much tape is required to stick the edges?
9. A farmer has 100 m of fence. He wants to build a square enclosure for his sheep. Find the area enclosed.
10. The farmer rearranges his 100 m of fence to a rectangle 10 m wide. What is the new area? What is the percentage decrease in area?

C Circumference

The perimeter of a circle is usually called the Circumference.

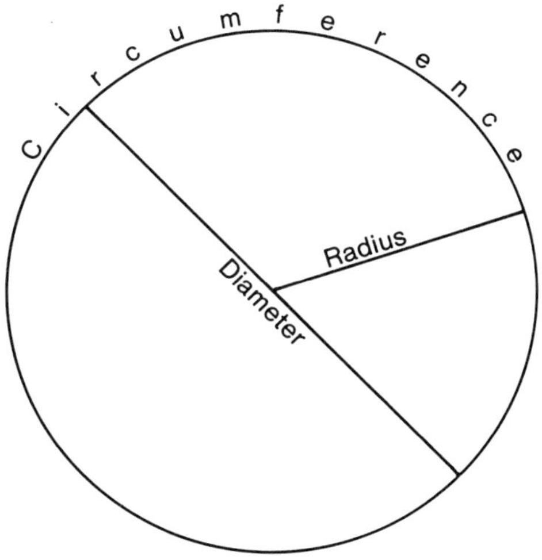

$C = \pi \times$ diameter

or $2 \times \pi \times$ radius

Remember π is about 3.14 or $\frac{22}{7}$

Learn:

Circumference $= 2 \times \pi \times r$

Example

Find the circumference of a circle radius 7 cm $\left(\pi = \frac{22}{7}\right)$.

$$C = 2 \times \pi \times r$$
$$= \frac{2}{1} \times \frac{22}{7} \times \frac{7}{1}$$
$$= 44 \text{ cm}$$

Draw a circle radius 7 cm: measure its circumference

Example

Find the circumference of a circle radius 2.1 cm (π = 3.14).

$C = 2 \times \pi \times r$

$2 \times 3.14 \times 2.1$

$= 6.28 \times 2.1$

$=$ **13.188 cm**

```
 6.28×
   2.1
 -----
   628
 12560
 -----
13.188
```

Fill in these examples:

1. *Find the circumference of a circle radius 14 cm* $\left(\pi = \frac{22}{7}\right)$.

   $C = 2 \times \pi \times r$

   $=$

   $=$ ANS

2. *Find the circumference of a circle radius 4.8 cm* ($\pi = 3.14$).

   $C =$

   $= 2 \times \quad \times$

   $=$ ANS

On paper find the circumference of these circles:

3. radius 21 cm $\left(\pi = \frac{22}{7}\right)$.
4. radius 1.25 cm ($\pi = 3.14$).
5. diameter 10 cm ($\pi = 3.14$).
6. diameter 7 cm $\left(\pi = \frac{22}{7}\right)$.

D Working Backwards

Example

Find the radius of a circle circumference 44 cm $\left(\pi = \frac{22}{7}\right)$.

$$C = 2 \times \pi \times r$$

$$r = C \div (2 \times \pi)$$

$$= 44 \div \left(2 \times \frac{22}{7}\right) = \frac{44}{1} \div \frac{44}{7}$$

$$= \frac{44}{1} \times \frac{7}{44} = 7 \text{ cm}$$

Either: learn how to rearrange the formula (divide both sides of equation by $2 \times \pi$)

or: learn $\boxed{r = C \div (2 \times \pi)}$

Example

Find the area of a circle with circumference 6.28 cm ($\pi = 3.14$).

$$r = \frac{C}{2 \times \pi} = \frac{6.28}{2 \times 3.14} = 1 \text{ cm}$$

$$A = \pi r^2$$

$$= 3.14 \times 1 \times 1$$

$$= \mathbf{3.14\ cm^2}$$

[You will be able to do harder examples with a calculator in Unit Fourteen.]

Try these on paper:

1. Find the radius of a circle circumference 88 cm $\left(\pi = \frac{22}{7}\right)$.
2. Find the circumference of a circle area 314 cm^2 ($\pi = 3.14$).

E Volumes

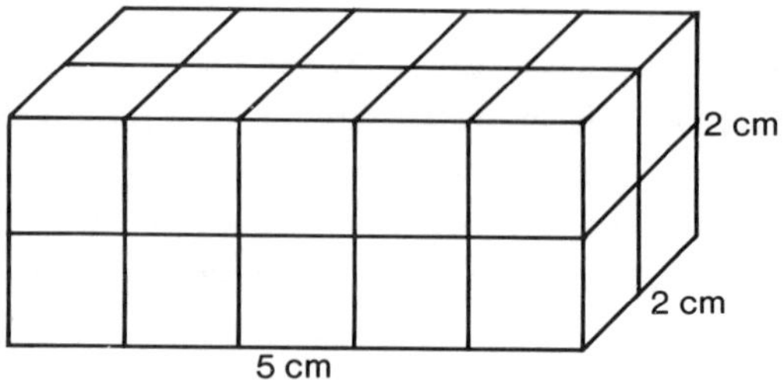

A unit of volume is cubic centimetre written cm^3.
The volume of a box is length × width × height.

> *Remember:*
> *Volume* $= l \times w \times h$

Example

Find the volume of a box length 4 cm, width 3.5 cm, height 20 mm.

Change all units to centimetres.

$$l = 4 \text{ cm}, w = 3.5 \text{ cm}, h = 2 \text{ cm}$$

$$V = l \times w \times h$$

$$= 4 \times 3.5 \times 2$$

$$= 14 \times 2$$

$$= \mathbf{28\ cm^3}$$

Fill in this example:

Find the volume of a box length 5 in, width 2.5 in, height 1.5 in.

$$V = l \times w \times h$$

$$= \quad \times \quad \times$$

$V =$ ANS

Try these questions. Fill in the gaps. Do the working on paper.
(Answer Q.2 in cubic centimetres, Q.5 in cubic inches.):

	Length	Width	Height	Volume
1.	5 cm	4 cm	3 cm	
2.	18 cm	90 mm	80 mm	
3.	3 cm	1 cm		6 cm^3
4.		3 m	6 m	72 m^3
5.	1 ft 3 in	9 in	1 ft 1 in	
6.	1.8 cm	1.6 cm	0.7 cm	

F Mixed Examples

Look back over this unit. Do you know the formulae? Try to fill them in:

$C=$

$r=$ (radius in terms of C)

$V=$ (volume of box)

Now try these mixed examples:

1. Calculate the circumference of a circle of diameter 21 cm $\left(\pi=\frac{22}{7}\right)$.
2. Find the area of a circle radius $3\frac{1}{2}$ in $\left(\pi=\frac{22}{7}\right)$.
3. A tank measures 225 cm long, 60 cm wide, 24 cm high. Find its volume.
4. Calculate the perimeter of a rectangle 18 cm long and 13 cm wide.
5. Find the area of a circle, diameter 14 m $\left(\pi=\frac{22}{7}\right)$.
6. Find the perimeter of a square which has area 64 m^2.
7. A circular lawn 14 m in diameter is to be treated with "Lawnfood" costing 3 p per m^2. Find the cost $\left(\pi=\frac{22}{7}\right)$.
8. Find the vertical height of a triangle base 7 cm, area 112 cm^2.
9. A path is 102 m long, 90 cm wide and 10 cm thick. Find the volume in cubic metres. If 1 m^3 of concrete costs £9 find the cost of the path.
10. A square has a perimeter of 200 feet. Find its area in square yards.

Now go on to Unit Eleven Test.

UNIT ELEVEN **Test**

Set your work out neatly on lined paper. Put your name at the top.

Questions 1–3: Find the perimeter (or circumference) of these shapes:

1.

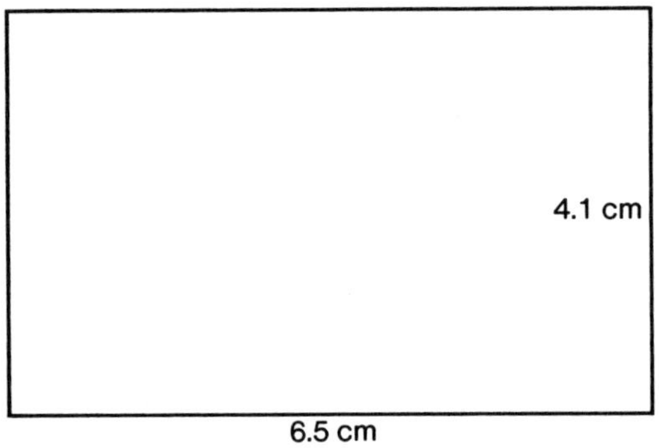

2.

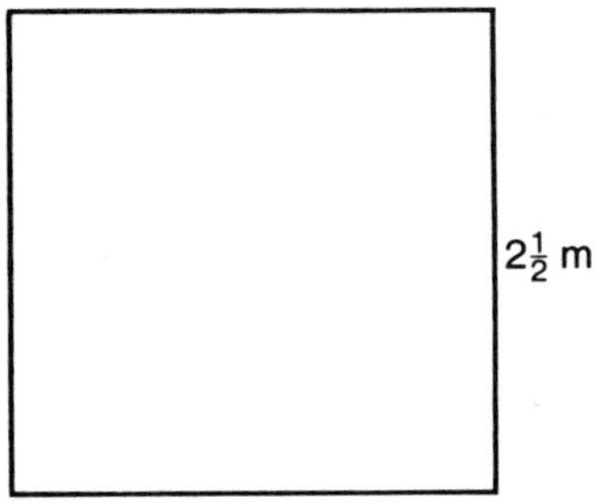

3.

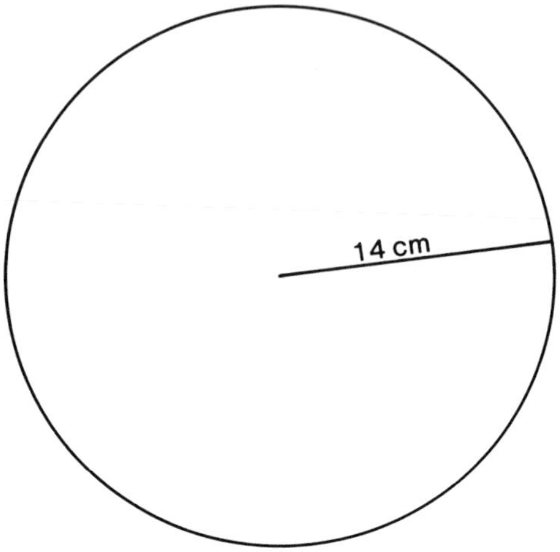

4. If the perimeter of a square is 11 cm find its area.

5. A circle has circumference 44 cm. Find its area $\left(\pi = \frac{22}{7}\right)$.

6. A machine uses a roll of paper 5 cm wide. If the roll contains 5 m^2 of paper how long is it?

7. The perimeter of a picture is 2 m. If it is 20 cm wide, how long is it?

8. Find the volume of a crate 50 cm wide, 1.5 m long and 50 cm high. (Answer in cubic metres.)

9. The circumference of a circle is 78.5 cm. Find the diameter ($\pi = 3.14$).

10. A cube has a volume of 8 cm^3. Find the length of the sides.

Do not forget to check your work.

UNIT ELEVEN ANSWERS

A 1. 24 cm 2. 12 cm

3. 4.8 m

B 1. $P = 17 \text{ cm} = 5 + l + l$
$l + l = 12 \text{ cm}$
$l = 6 \text{ cm}$
$A = l \times w = 6 \times 2.5$
$= \mathbf{15\ cm^2}$

2. $A = 25 \text{ cm}^2$
$l = 5 \text{ cm}$
$P = \mathbf{20\ cm}$

3. 16 m 4. 625 m^2

5. $3\frac{1}{2}$ cm, 16 cm

6. 0.49 m^2 = 4900 cm^2

7. 10 m; 9 m 8. 14.4 m^2; 15.4 m

9. 625 m^2 10. 400 m^2

10. cont. $\frac{^{9}\cancel{225}}{\cancel{625}_{\cancel{25}\,1}} \times \frac{^{4}\cancel{100}}{1} = \mathbf{36\%}$

C 1. $C = \frac{2}{1} \times \frac{22}{\cancel{7}} \times \frac{\cancel{14}^{2}}{1}$

$= 88 \text{ cm}$

2. $C = 2 \times 4.8 \times 3.14$
$= 30.144 \text{ cm}$

3. 132 cm 4. 7.85 cm

5. 31.4 cm 6. 22 cm

D 1. $r = \frac{C}{2 \times \pi} = \frac{88}{2 \times \frac{22}{7}}$

$= \frac{88}{1} \div \frac{44}{7}$

$= \frac{^{2}\cancel{88}}{1} \times \frac{7}{\cancel{44}}$

$= 14 \text{ cm}$

2. $\pi r^2 = 314$
$r^2 = 100$
$r = 10 \text{ cm}$
$C = 62.8 \text{ cm}$

E $V = 5 \times 2.5 \times 1.5 = 18.75 \text{ in}^3$

1. 60 cm^3

2. 1296 cm^3 or 1296 000 mm^3

3. 2 cm 4. 4 m 5. 1755 in^3

6. 2.016 cm^3

F $C = 2 \times \pi \times r$; $r = \frac{C}{2 \times \pi}$; $V = l \times w \times h$

1. 66 cm 2. $38\frac{1}{2}$ in^2

3. 324 000 cm^3 4. 62 cm

5. 154 m^2 6. 32 m

7. £4.62 8. 32 cm

9. 9.18 m^3; £82.62

10. $277\frac{7}{9}$ or 277.78 yd^2

UNIT TWELVE

Time

A The 24 hour clock

There are two methods in common use for stating the time: the 12 hour system and the 24 hour system.

Example

24 hour system		*12 hour system*
0000	=	12 midnight
0100	=	1.00 a.m.
0525	=	5.25 a.m.
1200	=	12 noon
1215	=	12.15 p.m.
1425	=	2.25 p.m.
1951	=	7.51 p.m.
2359	=	11.59 p.m.

Compare the two systems.

a.m. means before noon.
p.m. means after noon.

The . in 5.25 a.m. is not a decimal point: it divides the hours from the minutes. It means $5\frac{25}{60}$ hours past midnight.

Remember 60 minutes = 1 hour.
Try the following yourself:

1. Change to 24 hour clock:
   - a) 4.20 p.m.
   - b) 5.55 a.m.
   - c) 11.15 p.m.
   - d) 7.42 p.m.
   - e) 12.45 p.m.
   - f) 9.35 a.m.
   - g) 12 midnight
   - h) 12 noon
   - i) 1.05 p.m.
   - j) half past nine in the morning
   - k) quarter to five in the afternoon
   - l) 3 minutes to midnight

2. Change to 12 hour clock:
   - a) 0500
   - b) 1220
   - c) 2340
   - d) 0150
   - e) 2000
   - f) 1700
   - g) 0250
   - h) 1245
   - i) 0040
   - j) 1022
   - k) 0225
   - l) 1555

B Journey Times

Example

If a train leaves Kings Cross at 1035 and arrives at Doncaster at 1221, then how long is the journey?

h	m	
$^{11}\not{12}$	$^{60+}21-$	*Borrow 60 minutes 60 + 21 = 81*
10	35	*81 − 35 = 46* *Write the answer in hours and*
1	46	*minutes.*

The train takes **1 h 46 m.**
(NB 1.46 is *wrong* because it looks like .46 hours not 46 mins.)

Example

A man leaves London at half past eight on Monday morning and gets to Australia at twenty past nine on Wednesday evening. How long does his journey take? (All are London times.)

$^{23}\not{2}\not{4}$	$^{60}00-$	15 h	30 m +	on Monday
8	30	24 h	0 m	on Tuesday
15	30	21 h	20 m	on Wednesday
		60 h	50 m	

This journey takes **60 h 50 m** ANS

Fill in these answers:

1. *A train leaves at 1635 and arrives at midnight. How long is the journey?*

24	00–	(*Write midnight as 2400 in order to subtract.*)
16	35	

2. *A man is in hospital from 10.30 a.m. on Monday until 4.30 p.m. on Thursday. How many hours does he stay?*

24	00–	h	m on Monday
10	30	h	m on Tuesday
		h	m on Wednesday
		h	m on Thursday

He stays for ANS

Try these on paper:

Work out the length of time from:

3. 1430 to 2355
4. 0954 to 1643
5. Noon to 1753
6. 7.50 a.m. to 9.55 p.m.
7. 1230 to 2219
8. 1422 Monday to 2356 Tuesday.

C Speed

The average speed for a journey $= \dfrac{\text{total distance}}{\text{total time}}$

Units

If distance is in metres and time in seconds then the speed will be in metres/second (m/s).
Any combination of distance and time units could be used but the most useful are:

km.p.h. or km/h	kilometres per hour
m.p.h.	miles per hour
m.p.s. or m/s	metres per second

Remember

$$Speed = \frac{Distance}{Time} \qquad Time = \frac{Distance}{Speed}$$

$$Distance = Speed \times Time$$

A guide to memory.

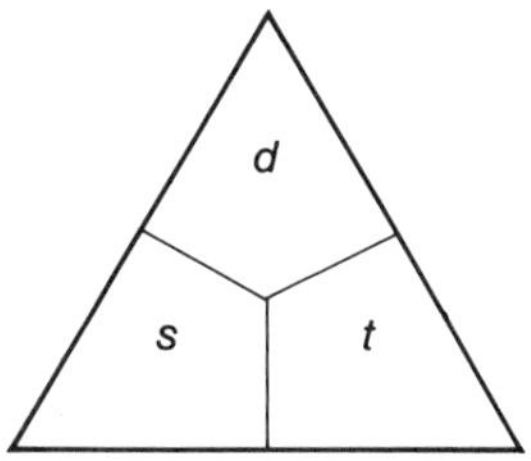

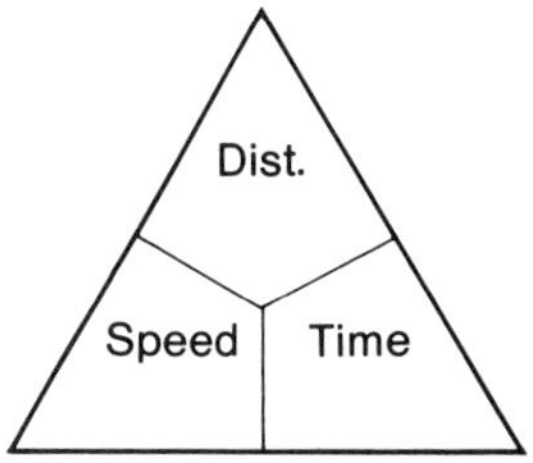

The letters are in alphabetical order.

Can you write down the formulae from the triangle?

Example

A car travels 100 km in 4 hours. Find the average speed.

Speed $= \dfrac{\text{distance}}{\text{time}}$ *Write down the formula.*

$= \dfrac{100}{4}$ km/h *Fill in the numbers. Think about the units!*

$=$ **25 km/h** ANS

Calculate the missing items:

	Distance	Time	Speed
1.	80 km	4 hrs	km/h
2.	km	4 hrs	200 km/h
3.	25 km	mins	50 km/h
4.	144 km		12 km/h
5.		$\frac{3}{4}$ hr	52 m.p.h.
6.	$2\frac{1}{2}$ miles		10 m.p.h.

D Timetables

Look at the railway timetable and answer the questions:

London to Humberside, Yorkshire, North East England and Scotland

		O FSO	E	J				O	1	O		H	SX		A		B			O	O	SX	f
London King's Cross 25 d	08 00					08 05	08 35		09 00			09 35	09 50	10 00				10 05		10 35	11 00		11.35
Stevenage 25 d	08u21					08 26	08 57											09 15					
Huntingdon 25 d	08 19					08 50												10 47					
Peterborough 25 a						09 08	09 30		09 51									11 06		11 27			
d	08u56					09 13	09 32	09 32	09 52									11 08		11 28	11u52		
Grantham a						09 33		10 03										11 28					
d						09 34		10 08										11 30					
Newark North Gate a						09 45												11 41					
30 Lincoln St. Marks a						10 10												12 30					
29, 30 Grimsby Town a						11 19						13x05						13 45					15i11
Newark North Gate d						09 47												11 42					
Retford a						10 01												11 57					
d						10 03												11 58					
Doncaster a	09 49					10 21	10 26		10 45			11 14						12 17		12 21			13 14
34 Wakefield Westgate a							10 47					11 58	11 45							12 42			13g59
34 Huddersfield a							11 38					12 35	12y49							13 42			15b25
34 Leeds a							11 06					12 18	12 03							13 02			14g19
40 New Pudsey a							11 47					12 47	12 27							13 47			14g47
40 Bradford Exchange a							11 57					12 57	12 39							13 57			14g57
35 Harrogate a	10v59						11 51		11v57			12v51								13 40			14v51
29 Hull a						11 18			12 03			12 38						13 28		13 23			14 29
29 Hull (via Selby) a					09 44						09 52							11 48					
Doncaster a	09 51								10 47		10 59	11 15		...				12 18					13 15
Selby d					10 23						11 21							12 35					
York a					10 57				11 14		11 38	11 43						12 52			13 09		13 43
39 Scarborough a	11 22								12 22			13 17									14 22		15 13
31 Sheffield d	09 08								09x49						11 22								
39 Leeds d	09 18		10 05						10 18			11 02			11 18		12 05				12 13		13 02
35 Harrogate d			09 39									10 51					11 42						12 46
39 Scarborough d									10 15												11 50		
York d	10 20		10 34						11 16			11 44			12 22		12 34				13 11		13 44
Thirsk d			10 57														12 57						
Northallerton d			11 08														13 08						
Darlington a			11 23						11 45			12 14			13 00		13 23				13 40		14 14
41 Middlesbrough 2 a									12 16			12 49			13 52						14 16		14 49
41 Stockton 2 a									12 19			13 19									14 19		
41 Hartlepool 2 a									12 44			13 44									14 44		
Darlington d		10 49	11 24						11 47			12 15			13 02		13 24				13 42		14 15
Durham a		11 09	11 46									12 34					13 46						14 34
d		11 11	11 48									12 35					13 48					14 05	14 35
Chester-le-Street d																						14 12	
Newcastle a	11 19	11 32	12 09						12 21			12 54		13 00	13 42		14 09				14 16	14 28	14 54
41 Sunderland 2 a	11 51	12 36	12 51						13 06			13 36		13 52	14 36		14 51				14 51		15 35
Newcastle 47 d	11 23								12 24					13 02							14 19		
Berwick-upon-Tweed 47 d									13 15												15 10		
Dunbar 47 d									13 44														
Drem 47 d																							
Edinburgh 47 a	13 02								14 12					14 43							16 03		
228 Glasgow Queen Street a	14 17								15 19					15 19							17 30		
241 Inverkeithing a	13 43								15 08					15 39							16 44		
241 Kirkcaldy a	14 01								15 31					16 01							17 06		
241 Cupar† a																							
241 Leuchars† a																							
241 Dundee a	14 49								16 19					16 49							17 55		
241 Arbroath a	15 29								16 38					17 30									
241 Montrose a	15 45								16 54					17 46									
241 Stonehaven† a																							
241 Aberdeen a	16 34								17 42					18 35									

For general notes see pages 2–4 of Timetable.

For complete service London King's Cross and intermediate stations to Peterborough see Table 25.

For local services Newcastle and intermediate stations to Edinburgh see Table 47.

† Through services only. For complete service from Edinburgh and intermediate stations to Aberdeen see Table 241.

• Stations for Rail Link coach service to or from Corby and Kettering see Table 26A.

A From Cardiff Central dep. 07 44 (Table 51).

B From Liverpool Lime Street dep. 10 05 (Table 39). Mondays to Fridays.

E From Birmingham New Street dep. 07 30 (Table 51).

H From Harwich, Parkeston Quay dep. 07 17 (Table 18) to Manchester Piccadilly arr. 12 48 (Table 29).

J From Liverpool Lime Street dep. 08 05 (Table 39).

b Saturdays only. Change at Leeds (Table 39).

g Saturdays only.

j Saturdays only. Change at Doncaster.

v Change at York.

y Change at Leeds (Table 39).

x Change at Doncaster.

(British Railways Board)

1. What time does the 0800 from Kings Cross get to Newcastle? How long does it take?
2. Find how long the 0805 takes to get to Hull.
3. Find how long the 0951 takes from Peterborough to York.
4. Find how long the 1143 takes from York to Newcastle.
5. Find how long the 0952 takes from Peterborough to Aberdeen.

Find a train and write down the time and length of journey from:

6. Grantham to Selby.
7. Stevenage to Sunderland.
8. Cardiff to Newcastle (see note).
9. Sheffield to Kirkcaldy.
10. You live in Peterborough and have to be in Newcastle for a meeting at 2 p.m. Which train would you catch? Describe briefly your journey (stops and times).

E Units

Example

A train travels at 125 km.p.h. How far does it travel in 45 mins?

$$\text{Speed} = \frac{\text{Distance}}{\text{Time}}$$

Speed = 125 km.p.h.

Time = 45 mins $= \frac{45}{60}$ hours.

Work in kilometres and hours.

distance = speed × time

$$\frac{125}{1} \times \frac{\cancel{45}^{3}}{\cancel{60}_{4}}$$

$$\frac{375}{4} = 93\tfrac{3}{4}\text{ km}$$

Example

Change 120 km.p.h. to m.p.s.

$$= \frac{\cancel{120}^{2\,1} \times \cancel{1000}}{\cancel{60} \times \cancel{60}^{3}}$$

Multiply by 100 to change kilometres to metres. Divide by (60 × 60) to change hours to seconds.

$$= \frac{100}{3}$$

$$= 33\tfrac{1}{3}\text{ m.p.s.}$$

Fill in these:

1. *An athlete averages 26 m/s. How far can he run in 4 minutes?*

   $d = s \times t$

   $= \quad \times \qquad$ m

   $= \qquad$ km

   s = 26 m/s

   t = 4 mins = $\qquad$ s

2. *Change 100 m/s to km/h.*

   $$= \frac{100 \times \qquad \times \qquad}{1000}$$

   = $\qquad$ ANS

Try these on paper:

3. A journey of 24 miles takes 40 minutes. Find the average speed in m.p.h.
4. A train travels 152 km in 1 h 16 m. Find the average speed.
5. Change 360 m.p.h. to feet per second.
6. Change 500 m/s to km/h.

F Wages and Overtime

Example

A clerk averages 15 hours a week for four weeks. He is paid £2.25 per hour. Find his wages for the month.

Hours worked = 15 × 4 = 60

$$\begin{array}{r} 15\times \\ 4 \\ \hline 60 \end{array}$$

Wages =

$$\begin{array}{r} 2.25\times \\ 60 \\ \hline 135.00 \end{array}$$

He earns **£135** ANS

Example

A man is paid £3.20 per hour for a 36 hour week, time and a quarter for overtime and time and a half for Saturdays. If he works 44 hours of which 3 hours is on Saturday how much will he be paid?

Basic wage = £3.20

Overtime @ $1\frac{1}{4} \times$ £3.20 = £4.00

Overtime @ $1\frac{1}{2} \times$ £3.20 = £4.80

44 hours − 3 hours on Saturday leaves 41 hours.

Overtime during the week = 41 − 36 = 5 hours.

Wages = 36 × £3.20 + 5 × £4.00 + 3 × £4.80

$$\begin{array}{r} 3.20\times \\ 36 \\ \hline 19\ 20 \\ 96\ 00 \\ \hline £115.20 \end{array} \qquad \begin{array}{r} 4.00\times \\ 5 \\ \hline £20.00 \end{array} \qquad \begin{array}{r} 4.80\times \\ 3 \\ \hline £14.40 \end{array}$$

Total Wage

$$\begin{array}{r} 115.20+ \\ 20.00 \\ 14.40 \\ \hline £149.60 \end{array}$$

He earns **£149.60** ANS

Set your work out to show the steps in the calculation. This will help you to check.

Try these questions yourself:

1. A girl is paid £1.90 per hour for a 30 hour week. Find the total pay.
2. A workman gets £3.00 per hour and double pay for Sundays. He works from 9 a.m. to 5 p.m. each day for seven days. Find his total pay.
3. A workman has a 36 hour week for which he is paid £72. He gets time and a half for overtime. How much is he paid for 40 hours?
4. A secretary works from 8.45 a.m. to 5.15 p.m. with 1 hour for lunch. She works four days and gets £1.50 per hour. Find her total pay.

G Mixed Examples

1. A flight was due to take off at 1620 but was delayed until 1813. How long was it delayed?
2. A man pays £500 annual rent and £150 rates. Find his weekly bill to the nearest penny (1 year = 52 weeks).
3. A woman invests £5 per month in Premium Bonds from July 1980 to August 1982. How much has she invested?
4. A car travels 256 miles at an average speed of 30 m.p.h. How long does the journey take?
5. A clock loses 2 seconds every day. How much time has it lost after one year. (1 year = 365 days, answer in minutes and seconds)
6. How many hours in $3\frac{3}{4}$ days?
7. A taxi charges 10 p for every 5 minutes plus 20 p per mile. A journey of 8 miles takes 15 minutes. How much does it cost?
8. A car travels 155 miles in $2\frac{1}{2}$ hours. Find the average speed.
9. An athlete completes 800 m in 3 minutes. Find his average speed in m/s.
10. Convert 80 m.p.h. to feet per second.

Now go on to Unit Twelve Test.

UNIT TWELVE **Test**

Set your work out neatly on lined paper. Put your name at the top.

1. Change "twenty to three in the afternoon", to the 24 hour clock.
2. A train leaves at 6.30 a.m. and arrives at 10.15 p.m. How long is the journey?
3. A plane is due to leave at 1115 but is delayed until 1321. How long is it delayed?
4. A boy travels $3\frac{1}{2}$ km to school in 25 minutes. Find his average speed in km.p.h.
5. A train is travelling at 80 m.p.h. How far would you expect it to travel in 5 minutes?
6. Change 20 km.p.h. to m.p.s.
7. A girl has five maths lessons a week. There are 40 school weeks in a year. If she wastes 5 minutes per lesson, how long does she waste in a year?
8. The Giant Tortoise has a recorded speed of 4.5 m per minute. How long would it take to travel 1 km?
9. A council worker earns £2.04 per hour for a 36 hour week and time and a half for overtime. How much does she earn if she works $41\frac{1}{2}$ hours in a given week?
10. A man earns £54 for a 36 hour week. He gets double time on Sundays and time and a half for other overtime. In a certain week he earns £66. He worked 2 hours on Sunday. What were his total hours worked?

Do not forget to check your work.

UNIT TWELVE ANSWERS

A

1. a) 1620 b) 0555 c) 2315
   d) 1942 e) 1245 f) 0935
   g) 0000 h) 1200 i) 1305
   j) 0930 k) 1645 l) 2357

2. a) 5.00 a.m. b) 12.20 p.m.
   c) 11.40 p.m. d) 1.50 a.m.
   e) 8.00 p.m. f) 5.00 p.m.
   g) 2.50 a.m. h) 12.45 p.m.
   i) 12.40 a.m. j) 10.22 a.m.
   k) 2.25 a.m. l) 3.55 p.m.

B

1.

2~~4~~ (3)	(60)00
−16	35
7 hr	25 m

2.

13	30	Monday
24	00	Tuesday
24	00	Wednesday
16	30	Thursday
78	00	It takes **78 h** ANS

3. 9 h 25 m 4. 6 h 49 m
5. 5 h 53 m 6. 14 h 5 m
7. 9 h 49 m 8. 33 h 34 m

C

1. 20 km/h 2. 800 km
3. 30 minutes 4. 12 h
5. 39 miles 6. quarter of an hour or 15 minutes

D

1. 1119, 3 h 19 m 2. 3 h 13 m
3. 1 h 23 m 4. 1 h 11 m
5. 7 h 50 m 6. 1130, 1 h 5 m
7. 0821, 3 h 30 m 8. 0744, 5 h 58 m
9. 0908, 4 h 53 m
10. Leave Peterborough 0952. Train stops at Doncaster 1045 for 2 minutes, stops at York at 1114 and Darlington 1145. Arrives at Newcastle 1221 with 1 h 39 m to spare.

E

1. 6240 m = 6.24 km 2. 360 km
3. 36 m.p.h. 4. 120 km/h
5. 528 feet per second 6. 1800 km/h

F

1. £57 2. £192 3. £84
4. £45

G

1. 1 h 53 m 2. £12.50
3. £130 4. 8 h 32 m
5. 12 m 10 sec 6. 90 hours
7. £1.90 8. 62 m.p.h.
9. $4\frac{4}{9}$ m/s or 4.44 m/s
10. $117\frac{1}{3}$ feet per second

UNIT THIRTEEN

Revision

You have now completed the basic skills part of this course. Revise the work so far before continuing.

Method of Revising

1. For each unit 1 to 12
   a) *Learn* any formulae and methods you have forgotten.
   b) *Repeat* questions that you originally found difficult.
2. Work through the tests in this section or some old exam papers if you have any.
   a) Allow 1 hour for each test.
   b) Work as you would in an examination, on your own without notes.
   c) Mark your work or ask someone to mark it for you.
   d) Don't leave it there! Now correct any you got wrong by looking at your previous notes. This step is most important.
   e) Record your mark then try another paper.
   f) Continue until you consistently get a good mark, at least 20 out of 30.
3. Keep revising.
   Every few weeks do a revision test or old exam paper, and you will see your marks steadily increase.

Paper One

UNITS 1 TO 3

1. $2000 - 252 =$
2. $10 \times 20 \times 30 =$
3. $182 \div 13 =$
4. $3.65 + 2.8 + 11.44 =$
5. $12\frac{1}{2} \times 9 =$
6. $5\frac{9}{16} - 3\frac{1}{4} =$
7. $1.92 \div 1.2 =$
8. $\frac{500}{112 + 188} =$
9. Take two thirds of 90 from one half of 180.
10. $2142 + 317 + 36 =$
11. $25\,088 \div 98 =$
12. $19\frac{2}{3} + 6\frac{3}{8} =$
13. $3 - 1\frac{7}{8} =$
14. $8\frac{1}{3} \div 2\frac{1}{2} =$
15. Change $\frac{3}{8}$ to a decimal.
16. $0.2 \times 0.35 =$
17. $3\frac{1}{4} \times 1\frac{1}{3} =$
18. $1\frac{1}{3} - \frac{5}{6} =$
19. $2\frac{1}{5} + 1\frac{1}{3} =$
20. $31.62 \div 100 =$

21. $119.73 - 82.6 =$

22. $0.8 \times 0.4 =$

23. Change 3.15 to a fraction.

24. Find $\frac{3}{4}$ of $4\frac{1}{2}$.

25. $2\frac{4}{5} - 1\frac{2}{3} =$

26. $2.835 \div 1.5 =$

27. $\frac{1}{8} \times \frac{3}{11} =$

28. $8\frac{1}{3} \div 1\frac{1}{9} =$

29. Change $\frac{5}{8}$ to a decimal.

30. $7.9 \times 0.16 =$

Paper Two

UNITS 1 TO 6

1. Change £6.84 to pence.
2. A book costs £3.99. If I buy two books, how much change do I get from £10?
3. Find the cost of 17 ℓ of petrol at 38 p per litre.
4. Find the cost of $2\frac{1}{2}$ kg of seed at £1.68 per kilogram.
5. Share £9.60 between five people so that they all receive the same amount.
6. Change 520 schillings to pounds if £1 = 26 Austrian schillings.
7. A man takes £200 to Germany. He spends 600 marks. How much has he left. Answer to the nearest pound. (£1 = 3.8 marks.)
8. Add 1.8 kg + 265 g + 1.95 kg + 1240 g = kg
9. Change 53 pints to gallons and pints.
10. How many 250 mℓ paper cups can be filled from 8 ℓ of beer?
11. Change 6 in to cm. (1 in = 2.54 cm.)
12. Change 25.4 cm to inches. (1 in = 2.54 cm.)
13. Arrange in order of size smallest first, 1.9, $1\frac{7}{9}$, 1.88.
14. $\sqrt{81} + \sqrt{36} =$
15. Find the HCF of 12, 18, 21.
16. $2\frac{1}{4} \times 1\frac{1}{3} =$
17. $3.6 \div 0.3 =$
18. $1\frac{5}{6} - \frac{8}{9} =$
19. $2000 - 64 =$
20. Convert 0.65 to a fraction and simplify.
21. Convert $\frac{7}{20}$ to a decimal.
22. A car averages 26 miles per gallon. The owner travels 10 000 miles per year. How much petrol does he use to the nearest gallon per year?
23. 8 litres ÷ 12 = (Answer in millilitres.)
24. Convert 6 gallons to litres. (1 gallon = 4.5 litres.)
25. Change 8.36 kg to kilograms and grams.
26. A man saves £5 each month from January 1982 to October 1984 inclusive. How much has he saved?
27. A dance ticket costs £1.50. The total money from all tickets sold amounts to £141. How many tickets were sold?
28. A florist uses 10 in of ribbon to make each bow. He buys 5 yards of ribbon. How many bows can he make with it?
29. A child has been saving 10 p coins. She has 53 coins. She spends £1.20 and 99 p on Christmas presents for her mother and father. She also wishes to buy presents for seven friends. What is the most she can spend on each friend's present?
30. A pint of milk costs 23 p. What is the price per litre to the nearest penny? (1 gallon = 4.5 litres.)

Paper Three

UNITS 1 TO 9

1. $315 + 21 + 1042 =$
2. $8096 - 981 =$
3. $9.76 + 0.08 + 16.1 =$
4. $8.32 \times 0.12 =$
5. $(0.4)^2 =$
6. $3\frac{1}{5} + 2\frac{3}{4} =$
7. $9\frac{1}{4} \div 3 =$
8. $3.62 \div 100 =$
9. $0.96 \times 1000 =$
10. 36 km $\div$ 5 $=$
11. Divide 120 in the ratio 5 : 3.
12. How many 20 pence pieces can I get for £10.80?
13. Find the total cost of 14 articles costing 55 p each.
14. A holiday costing £140 is reduced by 5%. Find the new price.
15. Find the average of 3.1, 3.4, 3.8, 4.1.
16. Change 65% to a fraction in its simplest form.
17. A television costs 2440 francs. Calculate the cost in pounds if £1 = 12.2 francs
18. Express $\frac{7}{8}$ as a percentage.
19. Find 65% of 4 m (answer in centimetres).
20. How many 15 fluid ounce bottles can be filled from 16 pt. (20 fluid ounces = 1 pint.)
21. A car travels 160 miles at an average speed of 25 m.p.h. How long does the journey take?
22. A man borrows £3000 to improve his house. He pays it back after 3 years with 8% Simple Interest. How much does he pay back altogether?
23. A bottle holds $1\frac{1}{2}$ ℓ of medicine. How many 5 mℓ spoonfuls does it contain?
24. A box contains 70 sweets, 14 are caramels. What percentage are caramels?
25. A girl has 80 p pocket money. She spends 25% on sweets and 17 p on a comic. How much is left?
26. Arrange in order of size, largest first, $7\frac{2}{5}$, $7\frac{1}{3}$, 7.45.
27. A jumper is knitted from twenty two 50 g balls of wool. What is the weight of the finished jumper in kilograms?
28. Two men share the profits of their company in the ratio 2 : 3. If the first man gets £1000 what are the total profits?
29. A minature railway charges 30 p for adults and 20 p for children. 15 adults and 22 children ride on the train. How much do they pay altogether?
30. A boy sells his camera for £27 making a loss of 25%. What did he pay for it?

Paper Four

1. $348 + 74 + 331 =$
2. $112.9 - 87.81 =$
3. $67 \times 76 =$
4. $54.8 \div 100 =$
5. $2\frac{1}{2} + 5\frac{3}{8} =$
6. $3\frac{1}{7} + 1\frac{3}{11} =$
7. $0.05 \times 0.12 =$
8. $0.008 \times 1000 =$
9. £15.12 $\times$ 6 $=$
10. £28.80 $\div$ 12 $=$
11. Change 2.765 kg to grams.
12. 2 m 53 cm $\times$ 14 $=$
13. A tank contains 40 ℓ of water. How many 350 mℓ cups can be filled from it?
14. Arrange in order of size, smallest first, $1\frac{3}{7}$, 1.37, 1.73.

15. Find 15% of £250.

16. A girl scores 18 out of 20 in a test. Express her score as a percentage.

17. Change $37\frac{1}{2}$% to a fraction in its lowest terms.

18. A man invests $500 for 3 years at 8.5% interest. Find the Simple Interest.

19. Find the average of 1.2 kg, 750 g, 1.865 kg, 500 g and 900 g.

20. A car travels for 2 h 20 m at an average speed of 30 m.p.h. How far does it travel?

21. Divide 75 p in the ratio 3 : 2.

22. The perimeter of a square is 28 cm. Find the area.

23. A plane leaves at 1220 and arrives at 1710. How long does the journey take?

24. Convert £250 to francs. (£1 = 10.4 francs.)

25. How many 10 p pieces can you get for a £5 note?

26. Complete the next two terms of the series: 11, 12, 14, 17, __, __

27. A boy saves 22 p every Saturday starting on July 3rd. How much has he saved by the end of August?

28. A library charges 15 p for every book borrowed. One day the takings are £46.50. How many books have been borrowed?

29. Find the area of a circle radius 7 cm. $\left(\text{Take } \pi = \frac{22}{7}.\right)$

30. In a factory ten men earn £97 per week and six men earn £86. Find the average wage to the nearest pound.

Paper Five

1. $1984 - 564 =$

2. $6.1 + 3.55 + 8.03 =$

3. $1287 \div 9 =$

4. $65.9 \times 100 =$

5. $7\frac{1}{3} - 4\frac{3}{4} =$

6. $8\frac{8}{9} \times 3\frac{3}{8} =$

7. $0.85 \div 0.5 =$

8. $875 \text{ kg} \div 1000 =$ (Answer in grams.)

9. £$1254 \div 8 =$

10. Find the cost of 15 packs of teabags at £1.44 per pack.

11. What is the square root of 100?

12. Simplify the ratio 36 : 48.

13. A tank contains 30 ℓ of water. How many 25 mℓ cups can be filled from it?

14. Arrange in order of size largest first, 23%, 0.26, $\frac{1}{4}$.

15. An article costing £240 is reduced by 15%. Find the new price.

16. 2000 coins each 2 cm in diameter are placed side by side in a line. How many metres long is the line?

17. Change $\frac{1}{5}$ to a percentage.

18. What is the area of a rectangle measuring 2.5 cm by 4.5 cm?

19. Find the average of 9 in, 1 ft 5 in, 11 in, 1 ft 4 in.

20. A car travels for 16 km on one litre of petrol. How much petrol does it need to travel 100 km? (Answer to the nearest litre.)

21. Divide £175 in the ratio 3 : 4.

22. A motorist travels 99 miles in $4\frac{1}{2}$ hours. Find his average speed.

23. A plane leaves at 1348 and arrives at 1854. How long does the journey take?

24. Convert £980 to dollars. (£1 = $1.30.)

25. A piece of wood is 2 m long. Two pieces of 24 cm are cut off it. The rest is cut into as many 5 cm pieces as possible. How much is left?

26. Complete the next two terms of the series: 4, 7, 10, 13, __, __.

27. 170 metal bars weigh 200 g each. If one kilogram of the metal is worth £33 find the total value.

28. Find the circumference of a circle radius $3\frac{1}{2}$ metres $\left(\pi = \frac{22}{7}\right)$.

29. Find the area of a circle radius 14 cm $\left(\pi = \frac{22}{7}\right)$.

30. A girl gets a library book on 22nd October for 4 weeks. By which date must she return it?

Paper Six

1. $512 - 298 =$
2. $62 \times 14 =$
3. $171 \div 19 =$
4. $7\frac{3}{4} - 4\frac{7}{8}$
5. $7\frac{1}{3} \div 3\frac{2}{3}$
6. £2.17 × 7 =
7. £5.80 ÷ 4 =
8. 4 m ÷ 5 = (Answer in centimetres.)
9. 2 kg 120 g × 10 =
10. Find 35% of £7.
11. Find the average of 9, 11, 13, 24 and 28.
12. How many 3 in pieces of cord can be cut from one yard?
13. Convert £13.50 into francs if £1 = 11.2 francs.
14. Divide £175 in the ratio 4 : 3.
15. Express a quarter of a kilogram in grams.
16. Write 5.0217 correct to 3 significant figures.
17. What is the total cost of 27 books at £1.05 each?
18. Write in order of size, smallest first: 76%, $\frac{3}{4}$, 0.74.
19. Find the perimeter of a square of side 7.2 cm.
20. An article priced at £15 is reduced by 20% in a sale. What is the sale price?
21. A train leaves London at 1333 and arrives at Sheffield at 1704. How long does the journey take?
22. A coat costs £35 without VAT. VAT is 15%. How much does the coat cost when VAT is added?
23. Ball bearings weigh 25 g each. How many do I need to weigh 5 kg?
24. Find the circumference of a circle of radius 14 cm $\left(\pi = \frac{22}{7}\right)$.
25. A room measures 4 m by 3 m. If carpet costs £7.99 per square metre how much will it cost to carpet the room?
26. A man rents a room from December 9th 1984 to February 20th 1985. The cost is £15 per night. How much does he pay?
27. A driver uses 15 ℓ of petrol per day. His van travels 12 km per litre. He drives 6 days per week. How many kilometres does he travel per week?
28. A man borrows £500. He has to pay 5% interest at the end of each month. How much interest will he have paid if he repays the money after 2 months?
29. A student goes on a journey of 328 km by train. The journey takes 2 hours 40 min. What is the average speed of the train?
30. A pudding takes 40 mins to prepare and needs to be cooked for $2\frac{1}{2}$ hours. The cook wishes to serve it at 8 p.m. At what time must she start to prepare it?

UNIT FOURTEEN

Calculators

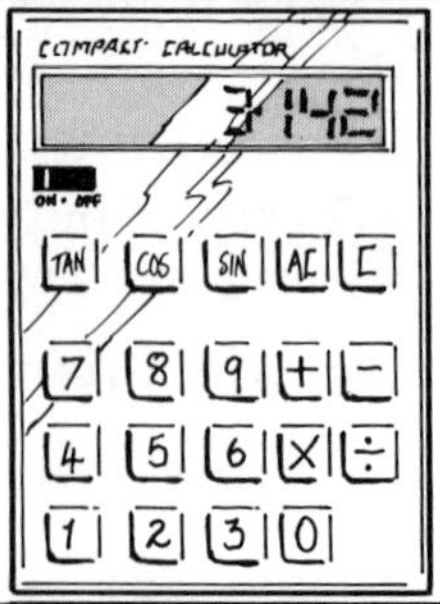

In Paper II of the examination you are allowed to use a calculator but you must show all the steps in your calculation.

Example

A family of two adults and two children take a 14 day holiday in a tent. The campsite charges £1.10 per night for adults and 40 p per child. 15% VAT is added to the bill. The family use 24.5 gall of petrol at £1.86 per gallon. Find the total cost of campsite and petrol.

Cost per adult = **£1.10 × 14 = £15.40**

Cost per child = **40 p × 14 = £5.60**

Cost for family = **2 × £15.40 + 2 × £5.60**

= **£30.80 + £11.20**

= **£42.00**

$$\textit{VAT} = \pounds\overset{21}{\cancel{42}} \times \frac{{}^{3}\cancel{15}}{\underset{10}{\cancel{100}}} = \frac{63}{10} = \pounds 6.30$$

Petrol cost = **£1.86 × 24.5 = £45.57**

Total cost = **£42.00 + £6.30 + £45.57**

= **£93.87**

The total cost is £93.87 ANS

Get into the habit of setting out all the stages of your calculation: it is easier for you to check it. Also other people can follow your method.

A Spot Your Errors

A calculator does not make you a mathematician. You need to know what calculations to perform and to be able to do them accurately.

Remember a calculator will give a wrong answer if you press the wrong buttons. Always have an idea of the size of the answer expected so you can quickly spot any errors you have made.

Example

Do a rough check on these calculations.

3.92 × 8.7 = 34.104

This is about 4 × 9 = 36 so the answer is probably correct.

65.96 ÷ 6.8 = 0.97

This is about 70 ÷ 7 = 10 so the answer is wrong. (It is 9.7.)

Spot the wrong answers in these calculations without using a calculator or doing calculations:

1. 9.62 × 8.14 = 78.3
2. 136 × .01 = 13.6
3. 24.9 ÷ 17.2 = 144.8
4. 35 ÷ 64 = 0.55
5. 1.75 × 0.05 = 0.0875
6. £3.12 ÷ $7\frac{1}{2}$ = £0.416
7. £62 × 21 = £302
8. £504 ÷ 1.65 = £305.45
9. 2.54 m × 18 = 45.72 m
10. 1.78 kg ÷ 203 = 361.34 kg

Correct wrong answers using your calculator.

B Calculations

Practice using your calculator correctly by doing these examples.
Estimate the approximate answer to check your calculation.
Check your answers before continuing:

1. $\frac{3.9 \times 0.08}{7.3}$ to 3 sig. fig.
2. Multiply 14.72 × 0.793 and give your answer to 3 sig. fig.
3. Find the value of $\frac{3.8 \times 7.9}{6.1 \times 2.8}$ to 4 sig. fig.
4. Change $\frac{11}{15}$ to a decimal correct to 3 decimal places.
5. Calculate $\frac{1}{2.1} + \frac{1}{7.2}$ correct to 3 sig. fig.

C The Metric System

You studied the Metric System in Unit Five. Look back to Unit Five if you have difficulty with these questions:

1. A rope is 2 m long. Two pieces are cut off, one 63 cm long the other 48 cm. How much rope is left?
2. A cyclist travels 2.8 km to work, then 800 m to a café for lunch. He returns to work after lunch via the post office increasing the distance travelled to 900 m. At 5 o'clock he cycles home. How far has he travelled during the day?
3. A roll of material is 21 m long. The shopkeeper cuts it into 2.5 m lengths. How many whole lengths does she cut?
4. A room measures 12.2 m × 14.1 m. A woman chooses a carpet 90 cm wide and decides to lay strips down the length of the room. How many metres of carpet must she buy to cover the room?
5. Three walls are to be covered with wallpaper and the window wall painted.

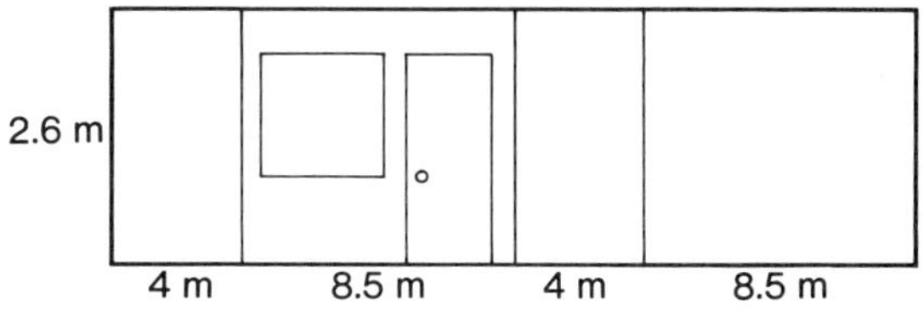

The wallpaper is 80 cm wide and each roll is 11 m long. How many rolls of wallpaper are needed?

(*Hint*: Wallpaper is stuck on the wall in vertical strips. How many strips will you need?)

D Money

Here are some more money questions to practise using your calculator.
Look back to Unit Four if you need help. Set out your work clearly:

1. Add £2.14 + £7.26 + £19.20 + £11 =
2. £6.92 × 23 =
3. A man earns £2.28 per hour. He gets time and a half for overtime. If he works 36 hours plus 8 hours overtime, how much will he be paid?
4. Find the cost of 95 units of electricity at 4.2 p per unit.
5. If 125 articles cost £581.25, how much are they each?

E Ratio and Proportion

Look back to Unit Nine if you have forgotten how to do these:

1. A road between two villages costs £9140 to repair. If the villages share the cost in the ratio of their populations 360 : 211, how much will each village pay? (Answer to nearest penny.)

2. Five organisations, Scouts, Cubs, Brownies, Guides and Rangers contribute towards a new campsite in the ratio 3 : 2 : 1 : 3 : 5. The Rangers contributed £95. What was the total money raised?
3. 12 eggs cost 75 p. What will 25 eggs cost? (Answer to nearest penny).
4. A car travels 108 km and uses one litre of petrol for every 9.4 km. Petrol costs 36 p per litre. Find the cost of the journey to nearest penny.
5. Change 916 francs to pounds if £1 = 9.81 francs. (Answer to nearest penny).

F Time

You have already met Speed $= \frac{\text{Distance}}{\text{Time}}$ in Unit Twelve.

If a journey is divided into two parts it helps to draw a diagram and show the speed, distance and time for each part. The average speed for the whole journey is $\frac{\text{Total Distance}}{\text{Total Time}}$.

Example

A man travels from London to Peterborough, a distance of 80 miles by coach, in 2 hours 40 minutes. He returns by train at an average speed of 70 m.p.h. Find

a) The average speed of the coach.

b) The time taken by the train.

c) The average speed for the return trip.

For each part of the journey write down all you know, speed, distance or time.

a) $Speed = \frac{Distance}{Time} = \frac{80}{2\text{ h } 40\text{ m}}$

We need to change the 2 h 40 m into a decimal. 40 minutes $= \frac{40}{60}$ *hours = 0.6666 . . . hours.*

So, 2 h 40 m = 2.6666 . . . hours.

The method on a calculator is

minutes ÷ 60 + hours.

Try this for 2 h 40 m and you should get 2.6666 . . .

Now back to the question.

$$Speed = \frac{Distance}{Time} = \frac{80}{2.6666\ldots} = 30 \text{ m.p.h.}$$

b) $Time = \frac{Distance}{Speed} = \frac{80}{70}$

$= 1.142\,857\ldots$ hours

This is 1 hour and a bit more. Write down the 1 hour and subtract it from the 1.142 857 . . . already on the calculator. This leaves 0.142 857 . . . hours. Multiply this by 60 to change to minutes.

time 1.142 857 . . . hours
= **1 h 9 m** to the nearest minute ANS

c) $Average\ Speed = \frac{Total\ Distance}{Total\ Time}$

$= \frac{80 + 80}{2\text{ h } 40\text{ m} + 1\text{ h } 9\text{ m}}$

$\frac{160}{3\text{ h } 49\text{ m}} = \frac{160}{3.8166\ldots}$

$= 41.92\ldots$

$=$ **42 m.p.h.** (to nearest mile per hour) ANS

Now fill in this example:

1. *A man travelled 150 miles by car in 4 h 30 m including a 25 minute coffee stop. Before his coffee break he averaged 45 m.p.h. Find his average speed for the second part of the journey.*

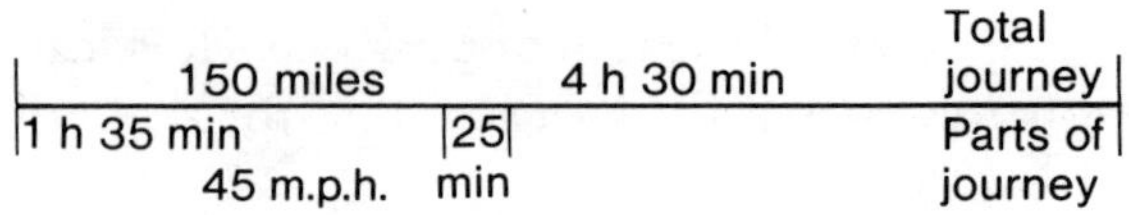

Before the break

$$\textit{Distance} = \textit{Speed} \times \textit{Time}$$
$$= 45 \times 1 \text{ h } 35 \text{ m}$$
$$= 45 \times 1.5833\ldots$$
$$=$$

(*Don't round off your answer until the end.*)

Distance left to travel = 150 – = miles

Time for the second stage

4 h 30 m – 1 h 35 m – 25 m =

$$\textit{Speed for the second stage} = \frac{\textit{Distance}}{\textit{Time}}$$
$$=$$
$$= \quad \text{m.p.h.}$$

Now try these questions:

2. A woman drives 5 miles through town at an average speed of 20 m.p.h. and then 65 miles on the motorway at 60 m.p.h. Find her average speed for the whole journey.

3. A train should complete a journey in 2 h 5 m but is delayed for 15 m. If the average speed is normally 80 m.p.h. find the average speed when it was delayed.

4. A boy cycles from school to his friend's house, a distance of 4 miles at 10 m.p.h. He stops there for 15 minutes and leaves his bike. He then walks the remaining 2 miles home at $3\frac{1}{2}$ m.p.h. If he left school at 4.00 p.m. at what time did he arrive home?

5. A girl decides to visit her grandmother who lives 53 miles away. She is taken to the station by car and catches the 0903 train which arrives at 1012. The guard tells her the train has averaged 43 m.p.h. She then walks one mile to her grandmother's house. How far is the girl's house from the station? (Answer to 2 decimal places.)

UNIT FOURTEEN **Test**

Set your work out neatly on lined paper. Put your name at the top. You may use a calculator but show *all* your working. These are exam type questions. Each question should take 10–15 minutes, but spend longer if necessary.

1. A concrete mix is made up of cement, sand and ballast in the ratio 2 : 1 : 1. Cement costs £3.00 per 25 kg bag, sand £3.50 and ballast £3.75.
   a) What weight of each would there be in 50 kg of mixture?
   b) A 25 kg bag of mixture costs £4.00. How much is saved by mixing your own? (Answer to the nearest penny.)
   c) What is the percentage saving in mixing your own?
2. Before going to Majorca a man changes £215 to pesetas (£1 = 183.2 pesetas) and the bank charges 1% for their services. On his return he has 910 pesetas which he changes back at the same rate. The bank again charges 1%.
   a) How many pesetas does he take on holiday? (Answer to nearest peseta).
   b) How much English money does he get back on his return?
   c) How much does the bank get for the two transactions (Answer to nearest pound.)
3. A director arranges a journey from London to York. It is 252 miles each way.
   a) He leaves at 1600 and averages 48 m.p.h. When does he arrive?
   b) The company pays him 23 p per mile for expenses. How much does he claim for car expenses?
   c) The rail fare is £62. How much more than this is the car cost as a percentage?
4. An engineering firm employs 120 people in the factory and 14 in the office. The factory workers average £121 per week and the office workers £98.
   a) What is the total wage bill?
   b) The factory workers get a 9% rise and the office workers 8%. Find the new wage bill.
   c) Find the average wage for the whole group before and after the increase.
5. A washing machine is advertised at £220 cash or 25% deposit and 24 monthly payments of £10.40.
   a) How much is the total HP price?
   b) What percentage extra is the HP price?
   c) Another shop offers the same machine for 12 monthly payments of £22.20. By what percentage is this dearer or cheaper than the original HP offer?

Do not forget to check your answers.

UNIT FOURTEEN ANSWERS

A

1. Correct
2. Wrong 1.36
3. Wrong 1.448
4. Correct
5. Correct
6. Correct
7. Wrong £1302
8. Correct
9. Correct
10. Wrong .008 77 kg or 8.77 g

B

1. 0.0427 2. 11.7
3. 1.758 4. 0.733
5. 0.615

C

1. 89 cm 2. 7.3 km 3. 8
4. 14 strips × 14.1 m = 197.4 m
5. 21 lengths of paper 2.6 m long. Four strips per roll. You need six rolls of paper (5 rolls will do if you join one strip).

D

1. £39.60 2. £159.16
3. £109.44 4. £3.99
5. £4.65

E

1. £5762.52 : £3377.48
2. £266
3. £1.56 (to nearest penny)
4. £4.14 5. £93.37

F

1. distance 71.25 miles; distance left 78.75 miles; time 2 h 30 m: speed 31.5 m.p.h.
2. 52.5 m.p.h. 3. 71 m.p.h.
4. 5.13 p.m. 5. 2.55 miles

UNIT FIFTEEN

Profit and Loss

A Percentages

This unit assumes that Unit Seven (Percentages) has been completed. Go back and revise Unit Seven if you wish. Extra examples on percentages are given for you to practise using a calculator.
Work all answers to 3 sig. fig.:

1. Convert $\frac{9}{13}$ to a percentage.
2. Convert 0.815 to a percentage.
3. What is 18.2% of £5.28?
4. A girl scored 19 out of 30 in a test. Express this as a percentage.
5. A shop buys 480 calculators. If 4% are faulty, how many are in good condition?

B Profit and Loss

Some terms you will meet are:

(CP) Cost price = the cost to the shopkeeper.

(SP) Selling price = the cost to the customer.

Profit or Loss = the difference between SP and CP.

Each of these can be expressed as a percentage. Cost price is always 100%.

Sometimes the questions seem complicated at first so set the information out in a table like this:

	£	%
CP		**100**
Profit or Loss		
SP		

From the table it is easier to calculate the answer. Study the examples carefully.

Example

A shopkeeper buys a TV for £200 and sells it for £250. Find his percentage profit.

	£	%
CP	**200**	**100**
Profit	**(50)**	**?**
SP	**250**	

Fill in the information given.
Calculate profit = £50

$$\pounds 200 = 100\%$$

$$\pounds 1 = \frac{100}{200}\%$$

$$\pounds 50 = \frac{50}{1} \times \frac{100}{200}\%$$

The profit = 25% ANS

Note the method:

Fill in the items given in the question.
Fill in any others that you can find by addition or subtraction (circled in notes to help you).
Put a question mark for the item you wish to find.
Write down the line that is complete either

a) £200 = 100% or b) 100% = £200

If you are finding a percentage use a) (percentage at the end).

If you are finding money use b) (pound at the end).

Then divide to find one pound or one percent. Multiply by the required number.

Do the calculation either on paper or by calculator (you must show method).

Example

A shopkeeper sells a table for £150, making 20% profit. Find the cost price.

	£	%
CP	?	100
Profit		20
SP	150	(120)

$$120\% = £150$$

$$1\% = £\frac{150}{120}$$

$$100\% = £\frac{150}{120} \times \frac{100}{1}$$

The CP = £125 ANS

Example

Fill in the gaps:

A boy buys a bicycle for £10, spends £5 on materials to repair the bike and then sells it for £25. Find his percentage profit.

	£	%
CP	15	100
Profit	( )	?
SP	25	

£15 = 100%

£1 = ————

= —— × %

= ANS

Example

A man buys a car and after four years sells it for £2200. If he makes a loss of 55% find the cost price (to the nearest ten pounds).

	£	%
CP	?	100
Loss		55
SP	2200	(45)

45% = £2200

1% =

100% =

CP = ANS

Now try these on paper. Use a calculator when necessary:

---

1. A shop sells a table for £156. If the profit is 30% find the CP.
2. A man buys an article for £15 and sells it for £18. Find his percentage profit.
3. A man buys a suit for £83.70. The tailor has made 35% profit. Find the cost of materials.

4. A shopkeeper buys a book for £1.00 and sells it for £1.80. Find the percentage profit.
5. A dealer sells a car for £3500 having made 40% profit. What did he pay for the car?
6. A painter buys materials for £4.00 and paints a picture. She sells it for £25.00. Find her percentage profit.
7. A boy buys an old bicycle for £10, repairs it and sells it making £15 profit. Find his percentage profit and selling price.
8. A greengrocer buys 100 oranges for £10 and sells them at 14 p each. Find his percentage profit.
9. A man sells his car for £900 making a 20% loss. What did he pay for it?
10. A girl buys ten bottles of perfume. She sells them at £1.50 each making about 30% profit. What did she pay for the pack of ten bottles? (Answer to nearest ten pence.)

C VAT

If a percentage is added or subtracted to a price such as VAT (tax) or for a sale the same method can be used even though the accounts are not called cost price and selling price. The original amount is always 100%.

Example

A plumber costs a job at £17.50 and adds 15% VAT. How much does he charge the customer?

	£	%
Cost	**17.50**	**100**
VAT		**15**
Total	**?**	**115**

$$100\% = £17.50$$

$$1\% = \frac{£17.50}{100}$$

$$\text{Total} = 115\% = \frac{£17.50}{100} \times \frac{115}{1}$$

$$= £20.125$$

$$= £20.13 \text{ ANS}$$

(to nearest penny)

Example

Fill in the gaps:

A retailer pays £115.24 for goods after a trade discount of 20% has been deducted. Find the list price of the goods.

	£	%
CP	**?**	**100**
Discount		**20**
SP	**£115.24**	**80** (circled)

80% = £115.24

1% = ________

100% = ___ ×

The list price is £ ANS

Now try these examples on paper:
(All answers to the nearest penny.)

1. A holiday is priced at £210 per person including VAT at 15%. What is the price before VAT?
2. Goods are priced in a catalogue at £520.40. The company offers $12\frac{1}{2}$% discount to staff. How much does a member of staff save on these goods?
3. A shop offers $2\frac{1}{2}$% discount provided bills are settled promptly. How much discount is allowed on a bill of £15.44?

4. A company makes a 4% surcharge on a holiday costing £410 to cover extra fuel costs. Find the new price.
5. Goods marked £5.49 are reduced to £4.49 in a sale. Find the percentage saving.

D Two Percentages

Sometimes we need to follow one percentage by another.

Example

An item costs £20 to manufacture. The factory wants 25% profit and VAT of 15% must be added. Find the final price.

a) *Firstly the 25%:* $£20 \times \frac{125}{100} = £25$

(*Draw up a table to see why this is correct.*)

b) *Secondly the 15%:* $£25 \times \frac{115}{100} = £28.75$

The final price is **£28.75** ANS.

Express this as a percentage increase of £20. (Not 40% as you might guess.)

	£	**%**
CP	**20**	**100**
Profit	**8.75**	**?**
SP	**28.75**	

$£20 = 100\%$

$£1 = \frac{100}{20}$

$£8.75 = \frac{100}{20} \times 8.75$

$= \mathbf{43.75\%}$ ANS

Alternatively:

$$20 \times \frac{\cancel{125}^{12.5}}{\cancel{100}_{10}} \times \frac{\cancel{115}^{11.5}}{100} = 20 \times \frac{143.75}{100}$$

This shows a 43.75% increase.

This method can also be used if you don't know the starting price.

In answering questions use whichever method you find easiest.

Try these:

1. A shopkeeper buys a coat for £25 and prices it to make 30% profit. During the sale he offers 25% off the ticket price. How much does the coat now cost? (Answer to nearest penny.)
2. A restaurant adds 10% service charge and then 15% VAT. What is the total percentage increase to the price?
3. A retailer has a TV for sale listed at £420. She offers 10% off the list price. She then offers a customer a further 5% discount for cash. How much does the customer pay?
4. A man earns £15 for a job. He gets 10% bonus but pays 10% of the total in tax. How much does he receive?

UNIT FIFTEEN **Test**

Set your work out neatly on lined paper. Put your name at the top.

1. A company produces handbags and shopping bags. The handbags sell at £4.50 and the shopping bags £6.00. The company sells 1000 handbags and 800 shopping bags in a normal week.
   a) How much money does the company receive in a week?
   b) How much profit does the company make in a week if the production cost of a bag is 80% of its *selling* price?
   c) During a flu epidemic the company only produce 500 handbags and 600 shopping bags. Express the money received as a percentage of a normal week's takings.
2. A painter estimates that a job will need ten rolls of wallpaper at £2.99 a roll, four litres of paint at £3.25 a litre and £5.25 worth of paste, filler etc. He charges £5.20 per hour for labour and the job takes 21 hours.
   a) Find the total cost of labour and materials.
   b) He adds 25% for his profit and VAT is 15%. What is the final bill? (Answer to nearest pound.)
   c) He allows 10% discount before VAT for December jobs. How much would the final bill be in December?
3. An article is priced at £75. If this gives the shop a profit of 25%, calculate:
   a) the cost price,
   b) the profit if 14 are sold.
4. A retailer buys 24 radios for £600. He prices them at £35 but offers 10% off in the sale.
   a) What is the profit on a radio sold in the sale.
   b) Express this as a percentage of the cost price.

Do not forget to check your answers.

A

1. 69.2% 2. 81.5% 3. 96.1 p
4. 63.3% 5. 461

B

Example

£15 = 100%

$£1 = \frac{100}{15}$

$£10 = 10 \times \frac{100}{15} = 66\frac{2}{3}\%$ or **66.7%**

Example

45% = £2200

$\mathbf{1\% = \frac{2200}{45}}$

$100\% = 100 \times \frac{2200}{45}$

= £4888.89

= **£4890**

1. £120 2. 20% 3. £62
4. 80% 5. £2500 6. 525%
7. 150%, £25 8. 40%
9. £1125 10. £11.50

C

80% = £115.24

$1\% = \frac{115.24}{80}$

$100\% = \frac{115.24}{80} \times 100 =$ **£144.05**

1. £182.61 2. £65.05
3. 39 p 4. £426.40
5. 18.2%

D

1. £24.38 2. $26\frac{1}{2}\%$
3. £359.10 4. £14.85

UNIT SIXTEEN

Your Money

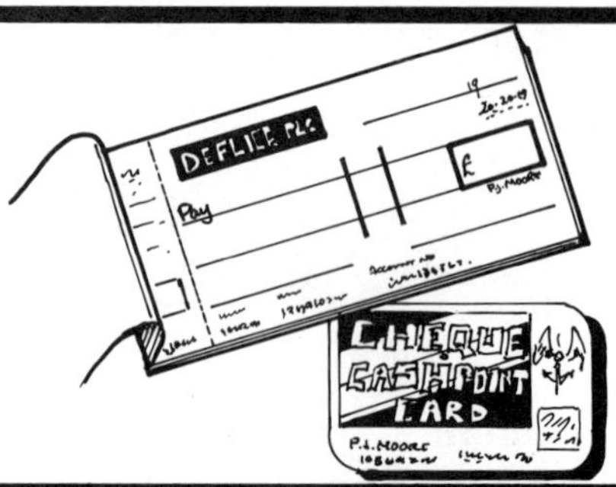

A Simple Interest

In Unit Seven you learnt to solve Simple Interest problems.

Example

Find the Simple Interest on £500 saved for 3 years at 5%.

Interest for 1 year is 5% of £500 $= \frac{5}{100} \times 500$

$= £25$

Interest for 3 years $= 3 \times £25 = £75$ ANS

A variation on this method is to use the Simple Interest Formula.

I = Simple Interest (pounds)
R = Rate of Interest percent per year
T = Time in years
P = Principal (pounds) (amount saved)

$$I = \frac{P \times R \times T}{100}$$

(make sure the units are the same i.e. R = rate, T = years, I = pounds, P = pounds)

The above example becomes:

$$I = \frac{£500 \times 5 \times 3}{100}$$

$$= £75$$

Fill in this example:

1. *Find the Simple Interest on £1500 invested for $2\frac{1}{2}$ years at 9%.*

$$I = \frac{P \times R \times T}{100}$$

$$= \frac{1500 \times 9 \times 2\frac{1}{2}}{100} = \qquad \text{ANS}$$

Sometimes we need to find the principal, rate or time.
The formula can be arranged:

$P = \frac{100 \times I}{R \times T}$	$R = \frac{100 \times I}{P \times T}$	$T = \frac{100 \times I}{P \times R}$

Learn these formulae.

Example

A man invests £200 for 4 years. He then has £300. Find the rate of interest.

$$P = £200$$
$$T = 4 \text{ years}$$
$$I = £300 - £200 = £100$$
$$R = \frac{100 \times I}{P \times T}$$
$$= \frac{100 \times 100}{200 \times 4}$$
$$= 12\frac{1}{2}\%$$

Fill in this example:

2. *A sum of money is invested for 3 years at $12\frac{1}{2}\%$. The Simple Interest is £30. Find the sum of money.* *The money invested is the principal* (*P*).

$$P = \frac{100 \times I}{R \times T}$$

$$= \frac{100 \times £30}{12\frac{1}{2} \times 3}$$

$$= £\qquad\qquad \text{ANS}$$

Now try these on paper:

---

3. A man invests £500 at 10%. How long will it be before he has £600? (His £500 plus £100 Simple Interest).
4. A boy invests £12 for 3 years and receives a total of £4.32 Simple Interest. Find the rate of interest.
5. A company invests £22 400 for 8 months at 15.75%. Find the Simple Interest. (Don't forget to change 8 months to years.)

Mark your work before continuing.

B Compound Interest

Here the interest is added to the principal at the end of each year and this also earns interest.

Example

Find the Compound Interest on £800 invested for 3 years at 10%.

Interest for the first year

$$= 800 \times \frac{10}{100} = £80$$

New principal at the end of the first year

$$= 800 + 80 = £880$$

Interest for the second year

$$= 880 \times \frac{10}{100} = £88$$

New principal at the end of the second year

$$= 880 + 88 = £968$$

Interest for the third year

$$= 968 \times \frac{10}{100} = £96.8$$

New principal at the end of the third year

$$= 968 + 96.8 = £1064.8$$

Total interest

$$= 1064.8 - 800 = £264.80$$

Fill in the gaps:

Find the amount after 3 years from £500 invested at 20% Compound Interest.

Interest for the first year

$$= 500 \times \frac{20}{100} =$$

New principal at the end of the first year

$$=$$

Interest for the second year

$$=$$

New principal at the end of the second year

$$=$$

Interest for the third year

$$=$$

New principal at the end of the third year

$$=$$

The amount after 3 years means the same as the new principal after 3 years

Amount = £ ANS

Now try these on paper:

Find the amount:

1. £600 for 2 years at 10% Compound Interest.
2. £400 for 3 years at 8% Compound Interest.
3. £1500 for 2 years at 12% Compound Interest.
4. £500 for 3 years at 4% Compound Interest.
5. £1000 for 4 years at 5% Compound Interest.

Find the Compound Interest:

6. £100 for 2 years at 10%.
7. £150 for 2 years at 5%. (Answer to nearest penny.)
8. £30 for 3 years at 2%.
9. £1000 for 4 years at 10%.
10. £1200 for 2 years at $12\frac{1}{2}$%.

C Depreciation

Depreciation (loss in value) can also be compounded.

Example

(Fill in the gaps):

A machine costs £12 000. It depreciates by 20% per year. Find its value after 2 years.

Depreciation for the first year

$$12\,000 \times \frac{20}{100} = £2400$$

Value after one year

$$= 12\,000 - 2400 =$$

Depreciation for the second year

$$= \quad \times \frac{20}{100} =$$

Value after 2 years

$$= \qquad = £ \qquad \text{ANS}$$

Try these questions on paper:

1. A car costs £6500. It depreciates by 10% in the first year and 8% in the second year. Find its value after 2 years.
2. A man has £5000. Each year he spends 15% of his remaining capital. How much remains after 3 years? (Answer to nearest penny.)
3. A machine costs £1800. It depreciates by 8% a year. Find its value after 2 years.

D Wages

The rest of this unit is about wages. You should already know how to calculate overtime rates. See Unit Twelve if you have forgotten.
Now study this example involving wages.

Example

A man earns £91.80 for a 36 hour week. He gets time and a quarter for overtime.

a) *How much is he paid if he works 40 hours?*
b) *If he earns £117.30 how many hours' overtime has he worked?*

a) *His basic pay per hour* $= \frac{£91.80}{36} = £2.55$

Overtime is $1\frac{1}{4} \times £2.55 = 1.25 \times 2.55$

$= £3.1875$

40 hours worked = 36 hours + 4 hours overtime.

36 hours pay = £91.80+

4 × £3.1875 = £12.75

Total Pay = **£104.55** ANS

b) *Overtime pay* = £117.30−

91.80

£25.50

Hours overtime $= \frac{25.5}{3.1875} =$ **8 hours** ANS

Now fill in these examples:

1. *A painter earns £98.80 for a 38 hour week. He gets time and a half for overtime on weekdays and double time for Sundays. How much will he earn for 45 hours, two of which were on Sunday?*

   Basic hourly wage $= \frac{£98.80}{38} =$

   Overtime on weekdays = × *1.5* =
   Overtime on Sunday = × *2* =

   Hours worked 38 normal plus *weekdays overtime plus 2 on Sunday* = *45*
   Wages
   38 hours @ = £98.80 +
   hours @ =
   2 hours @ =
   Total £ ANS

2. *A part-time cleaner is paid £25.20 for working 14 hours and time and a quarter for overtime. If he earns a total of £38.70 how many hours has he worked?*

   Basic hourly wage $= \frac{£25.20}{14} =$

   Overtime rate = × 1.25 =

   Overtime pay = £38.70 – £25.20 =

   Hours overtime = =

   Total hours worked = 14 + = ANS

   Solve these on paper:

---

3. A woman is paid £100.80 for a 42 hour week. She gets time and a half for overtime. What is she paid for 48 hours?
4. A factory pays £2.16 per hour for a 36 hour week, time and a quarter for the first 5 hours overtime and time and a half for further overtime. A man works 45 hours. How much is he paid?
5. A Babysitting Agency charges 60 p per hour before midnight and double after midnight. A couple go out at 2030 and pay £3.90. What time did they get home?

E Commission

Salespeople are often paid commission on goods they sell.

Example

A salesman gets a basic weekly wage of £85.00 and 5% commission. How much does he earn if he sells £200 worth of goods in a week?

Commission = 5% of £200 $= \frac{5}{100} \times \frac{200}{1}$

= £10 +

Basic wage = £85

Total Wage = **£95**

Example

A factory worker is paid £45 per week and 2 p for every product he makes over 2000. How much does he earn if he produces 5000 products?

Commission paid on 5000 – 2000 products
= 3000 products

3000 × 2 p = 6000 p = £60 +

Basic wage = £45

Total wage = **£105** ANS

Try these:

---

1. A salesman is paid on commission only. He receives 25% of the cost of goods sold. How much does he get for £300 worth of sales?
2. A shop assistant is paid £42 per week plus 2% commission. She sells £250 worth of goods. Find her total weekly wage.
3. A saleswoman is paid 5 p for each book she sells in a week plus £18.00. She sells 914 books. How much will she receive?

4. In 1983 a salesman sold £23 812 worth of goods. He received $2\frac{1}{2}$% commission plus a basic salary of £4200. He paid tax at 30% on all his salary. In 1984 his sales dropped by 5% but his basic salary increased by 10%. Find his total salary before and after tax for:
   a) 1983 b) 1984
5. A worker is expected to pack ten items an hour but gets paid a bonus of 15 p for every extra item she packs. A woman works 28 hours in a week and packs 325 items. How much bonus will she receive?

F Tax

Everyone earning money is liable to pay tax. A certain amount of money is allowed tax free, the rest is taxed at various rates.

In 1983/84 a single person was allowed £1785 tax free, and a married person £2795 tax free.

Any money paid to superannuation funds or National Insurance is tax free but other earnings are taxed.

Up to £14 600 is taxed at 30%.

£14 600–£17 200 is taxed at 40%.

£17 200–£21 800 is taxed at 45%.

£21 800–£28 900 is taxed at 50%.

£28 900–£36 000 is taxed at 55%.

£36 000 upwards is taxed at 60%.

Example

Calculate the tax paid by a single man earning £15 400 and paying 5% superannuation and £100 National Insurance.

Superannuation = 5% of £15 400

$= \frac{5}{100} \times \frac{15400}{1} = £770$

National Insurance = £100

Single Person's allowance = £1785

Total allowances = £2655

15 400

Taxable pay = 2 655

12 745

Tax paid = $£12\,745 \times \frac{30}{100}$

= **£3823.50**

Example

A married man earns £29 000. How much tax does he pay?

29 000–

Married man's allowance 2 795

Taxable pay = £26 205

Tax 14 600 × 30% = £4380

(17 200 − 14 600) × 40% = £1040

(21 800 − 17 200) × 45% = £2070

(26 205 − 21 800) × 50% = £2202.50

Tax paid = £9692.50

In an examination use the tax rates and deductions given in the question; in real life cases use up-to-date tax tables. Try these examples. Using the 1983/84 rates quoted opposite find the tax payable:

1. A single woman earning £4200 paying £120 National Insurance.
2. A married man earning £8000 paying 5% superannuation and £200 National Insurance.
3. A married person earning £21 000 after deductions for superannuation and insurance.

Exam Type Questions:

4. A married man earns £12 400. He pays 6% superannuation and £300 National Insurance. He pays tax at 30%. How much tax does he pay?
5. A single person with £1780 tax allowance earns £22 000. He pays tax as follows:
   £12 000 at 30%
   the next £5000 at 40%
   the rest at 50%
   Calculate the total tax payable.

UNIT SIXTEEN **Test**

Set your work out neatly on lined paper. Put your name at the top.

1. A man who has £3500 to invest for a period of 2 years has the following choice of investments:
   A At $13\frac{1}{2}$% per year Simple Interest where he has to pay tax at 30% on the interest.
   B In a Building Society at 9% per year Compound Interest. Tax is paid by the society.
   a) Find the amount after 2 years before tax when the money is invested in A.
   b) Find the amount in the Building Society after 2 years when invested in B.
   c) Which is the better investment after tax, and by how much?
2. A finance company pays Simple Interest at three rates depending on type of investment:
   (i) 9.75% per year
   (ii) 10.5% per year
   (iii) 11.25% per year
   a) Calculate the amount of Simple Interest paid (in cash) on an investment of £2500 at the end of one year for each rate.
   b) Express the cash difference between (i) and (iii) above as a percentage of (i). (Answer to 1 decimal place.)
   c) The company lends money at 19% per year Compound Interest for short term loans. What will the borrower have to pay back in total if the company lends him £4000 for 2 years?
3. A saleswoman gets £2000 per year salary plus 12% commission on sales. She pays 10% of her earnings to a pension fund and £56 per month insurance. If she sells £25 000 worth of goods in a year calculate her average monthly salary after paying pension fund and insurance.

Do not forget to check your answers.

A

1. £337.50
2. £80
3. 2 years
4. 12%
5. £2352

B

Example

Amount after 3 years = **£864**

1. £726
2. £503.88
3. £1881.6
4. £562.43
5. £1215.51
6. £21
7. £15.38
8. £1.84
9. £464.10
10. £318.75

C

12 000 − 2400 = 9600

Depreciation in 2nd year $= 9600 \times \frac{20}{100}$

= 1920

Value after 2 years = 9600 − 1920

= **£7680**

1. £5382
2. £3070.63
3. £1523.52

D

1. Basic hourly wage = £2.60. Overtime rates £3.90 and £5.20

38 hours @ £2.60 =	£98.80
5 hours @ £3.90 =	£19.50
2 hours @ £5.20 =	£10.40
Total wage	= **£128.70**

2. Basic wage = £1.80

   Overtime = £2.25

   Overtime £13.50 = 6 hours

   Total = **20 hours**

3. Basic rate £2.40, £122.40
4. £104.22
5. 0130

E

1. £75
2. £47
3. £63.70
4. a) £4795.30; £3356.71

   b) £5185.54; £3629.87
5. £6.75

F

1. £688.50
2. £1381.50
3. £7130
4. £2568.30
5. £7210

UNIT SEVENTEEN

Household Expenditure

A Rates

Every property has a rateable value fixed by the council. The amount of rates paid each year depends on the rateable value and the rate levied by the council.

Example

If the rateable value of a house is £200 and the council levy £1 in the pound rates, then the owner pays £200. Usually the rate levied is not exactly £1 in the pound.

Example

If the rate levied is 150 p in the pound the owner pays 150 p for each pound of the rateable value.

Rates paid = 200 × 150 p

= £300

Fill in the rate payable in each case:

	Rateable value	*Rate levied*	*Rate paid*
1.	£200	150 p	£300
2.	£150	160 p	
3.	£300	110 p	
4.	£250	105 p	

Check you have got these right before continuing. If you know the rate paid and rate levied you can calculate the rateable value.

Example

Taking the same example as before. If the owner pays £300 rates and the rates levied are 150 p in the pound find the rateable value of the house.

Rateable value $= \dfrac{300 \times 100}{150}$

(*Change £300 to pence and divide by 150 p*)

= **£200**

Similarly you can calculate rate levied.

Example

A house has a rateable value of £200. A man pays £300 rates. Find the rate levied.

Rate levied $= \dfrac{300}{200} \times 100$ p

= **150 p** in the pound.

Rate Payable (pounds) =

$$\frac{\text{Rateable Value (pounds)} \times \text{Rate Levied (pence)}}{100}$$

Rateable Value (pounds) =

$$\frac{\text{Rate Payable (pounds)} \times 100}{\text{Rate Levied (pence)}}$$

Rate Levied (pence) =

$$\frac{\text{Rate Payable (pounds)} \times 100}{\text{Rateable Value (pounds)}}$$

Fill in the gaps:

	Rateable value	Rate levied	Rate payable
5.		110 p	£550
6.		115 p	£170
7.	£200		£250
8.	£240		£324
9.	£160	115 p	
10.	£150		£157.50

Sometimes you have to work with total values for a whole town.

Example

The rateable value of a town is £160 000. The council needs £20 000 to repair a bridge. How much must they put on the rates to pay for it?

$$\text{Rate levied} = \frac{\text{Rate payable} \times 100}{\text{Rateable value}}$$

$$= \frac{20\,000 \times 100}{160\,000} \text{ p}$$

$$= 12.5 \text{ p in the pound}$$

Example

The rateable value of a town is £500 000. The council levy a 5.5 p in the pound rate for the fire brigade. How much do they give the fire brigade?

$$\text{Rate Payable} = \frac{\text{Rateable value} \times \text{rate levied}}{100}$$

$$= \frac{500\,000 \times 5.5}{100}$$

$$= £27\,500$$

Try these questions on paper:

11. A man's house has a rateable value of £282. In 1983 the rate levied is 116 p in the pound. In 1984 the man pays £332.76 rates. Find a) the rate payable in 1983, b) the rate levied in 1984, c) the increase as a percentage. (Answer to 1 decimal place.)

12. In 1981 Newborough City levied a rate of 130 p in the pound. Of this 14 p went to the Police Force, 3.4 p to the Fire Service and 75 p to Education. (These are simplified figures.) The estimated product of a penny rate is £236 000 for the city. Find:
    a) the total raised for the Police,
    b) the amount a family with a house rated at £300 has contributed to education,
    c) the total raised for the Fire Service,
    d) the total rates paid in the city.

B Rent and Mortgage

Many people either rent their home or borrow the money to buy it. If they borrow the money it is called a mortgage and they have to pay it back with interest.

Example

A couple pay £18 per week rent and £210 per year rates. Find the total monthly payments.

Rent per year	= £18 × 52 =	£936+
Rates per year	=	£210
		£1146

Monthly payments = £1146 ÷ 12
= **£95.50**

Try these examples:

1. A couple pay £90.12 per month mortgage repayments. The rateable value of their house is £210. The rate levied is 85 p in the pound. How much does their house cost per year?
2. A flat costs £10.50 per week for rent. The rates are £109.20 per half-year. Find the total weekly cost.
3. The rateable value of a house is £170. The rate levied is 115 p in the pound. The owner lets it at £10.50 per week including rates. How much per year has the owner left after he has paid the rates?

C Bills

When working out a bill set it out clearly.
Fill in these examples:

1. *A quarterly gas bill is made up as follows:*

   A standing charge of £9.90.

   The first 52 therms at 37 p per therm.

   Any remaining therms at 35 p per therm.

   Find the cost when 362 therms are used.

   Number of therms at **35 p = 362 − 52**

   Bill
   52 *therms @* **37 p =** +
   therms @ **35 p = £9.90**

   Total ANS

2. *A quarterly electricity bill is made up of a fixed charge of £6.43 and 5.238 p per unit used. Calculate the cost if 782 units are used. (Answer to nearest penny.)*

   782 units cost 782 × 5.238 p =

   Total cost = £6.43 + = ANS

Try these examples on paper. Answers to nearest penny:

3. A gas bill consists of a standing charge of £8.20 plus 18.2 p per therm. Find the total bill for a family using 92 therms.
4. A telephone bill consists of £16.45 standing charge plus 5 p per unit plus VAT at 15%. A man uses 251 units. Calculate his total bill.
5. A man travels 8000 miles in a year. His car averages 32 miles per gallon and petrol costs £1.80 per gallon. He spends £80 on new tyres, £110 on servicing and repairs and estimates his car depreciates by £800. Find:
   a) the total petrol cost per year,
   b) the total running costs per year (petrol, depreciation, repairs etc.),
   c) the total cost per mile,
   d) the total cost per week.
6. A man pays a £6.50 standing charge per quarter plus 4.15 p per unit of electricity used. He uses 1185 units in a quarter. Find his total bill to the nearest penny.
7. A gas board has two tariffs.
   Tariff A. Standing charge £7 Units 35.2 p
   Tariff B. No standing charge Units 45.1 p
   Mr. Smith uses 251 units in a quarter on Tariff B. Should he change to Tariff A? How much, if any, would he save?
   Mrs. Green uses 92 units on Tariff A. Should she change? How much would she save?
8. A man buys a car for £4300 and sells it eight months later for £3500. He travels 6400 km. He estimates he travels 8 km per litre of petrol. Petrol costs 36 p per litre. He spends £48 during the eight months on other items and repairs. Find:
   a) the cost of eight months' motoring,
   b) the average cost per kilometre. (Answer to nearest penny.)

UNIT SEVENTEEN **Test**

Set your work out neatly on lined paper. Put your name at the top.

1. A large house with a rateable value of £520 is shared by two families. They share the rates in the ratio 3 : 2. The rate levied is 130 p in the pound.
   a) What is the total rate for the house?
   b) How much does each family pay?
2. A telephone bill is made up of £15.60 rental plus 5 p per unit used and VAT at 15% of the total bill. If a household uses 117 units, how much will they pay?
3. Mr Ford buys a car for £3700 and sells it after a year for £2900. He spends £42 on new tyres, £156 on insurance and tax and uses the car to travel 8,000 miles. The car averages 28 miles per gallon of petrol. The petrol costs £1.72 per gallon. Find:
   a) the total cost of his year's motoring.
   b) the average total cost per mile,
   c) the percentage of the total cost spent on petrol.
4. The rateable value of a town is £2500 000.
   a) A rate of 1.54 p in the pound is levied for the hospital. Find the amount raised.
   b) £125 000 is needed for a new car park. Find the rate in the pound needed to pay for it.

Do not forget to check your work.

UNIT SEVENTEEN ANSWERS

A

2. £240
3. £330
4. £262.50
5. £500
6. £1147.83
7. 125 p or £1.25
8. 135 p
9. £184
10. 105 p
11. a) £327.12 b) 118 p c) 1.7%
12. a) 14 × 236 000 = £3304 000
    b) 300 × 75 p = £225
    c) 3.4 p × 236 000 = 802 400 p = £8024
    d) 130 p × 236 000 = 30 680 000 p = £306 800

B

1. £1259.94
2. £14.70
3. £350.50

C

1. 52 therms at 37 p = £19.24
   310 therms at 35 p = £108.50
   £9.90
   £137.64
2. 782 × 5.238 p = 4096.116 p
   = £40.96 +
   £5.43
   £46.39
3. £24.94
4. £33.35
5. a) £450 b) £1440
   c) 18 p d) £27.69
6. £55.68
7. Mr. Smith. Tariff B = £113.20
   Tariff A = £95.35 **should change, would save £17.85** ANS
   Mrs. Green. Tariff A = £39.38
   Tariff B = £41.49 **should not change** ANS
8. a) £1136 b) 18 p

UNIT EIGHTEEN

Boxes, Cans and Prisms

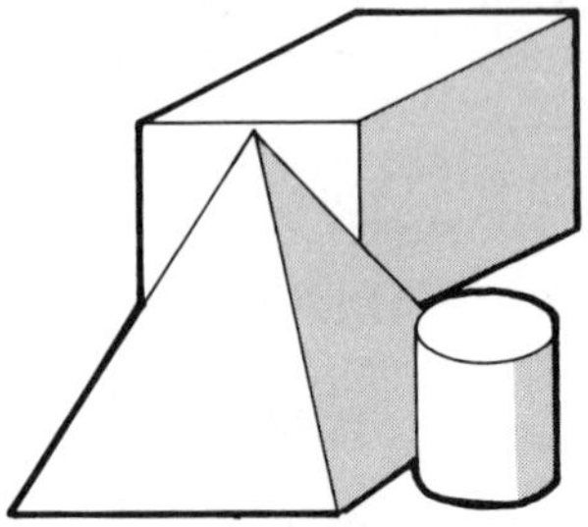

A Pythagoras' Theorem

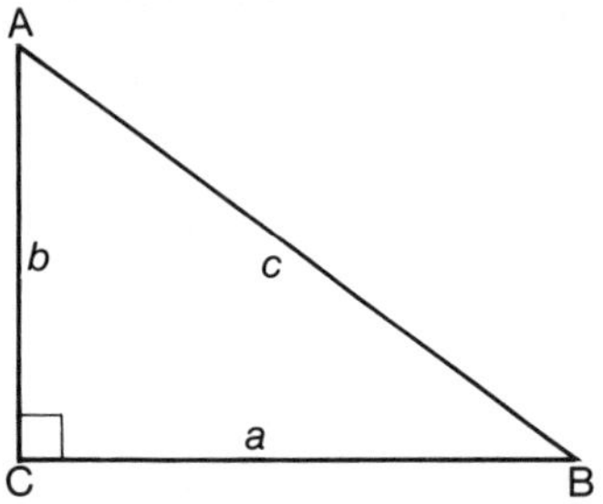

Measure the sides in centimetres AC = cm

BC = cm

AB = cm

Now square each measurement

$AB^2 =$ $\quad AC^2 =$ +

$BC^2 =$

$AC^2 + BC^2 =$

You should find $AC^2 + BC^2 = AB^2$

You can call the sides a, b, c then $a^2 + b^2 = c^2$:

Draw another right angled triangle and check that $a^2 + b^2 = c^2$ (c must be the longest side)

Fill in these examples:

1.

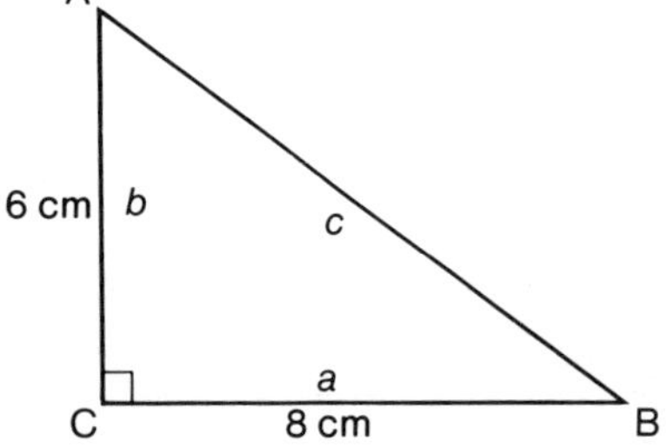

Find c:

$a^2 + b^2 = c^2$

$8^2 + 6^2 = c^2$

$64 + \quad = c^2$

$c^2 =$

$c =$ cm

2.

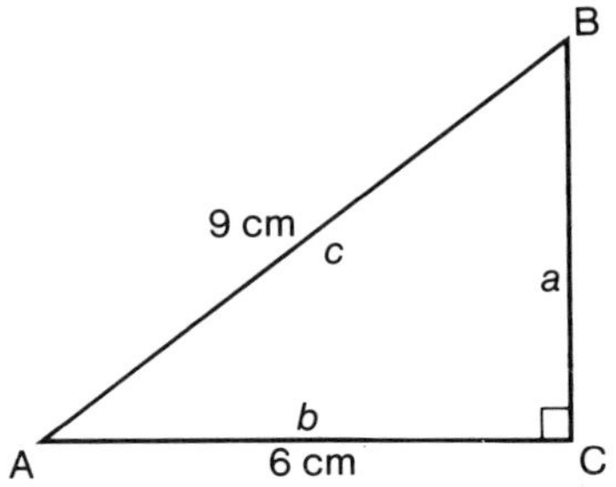

Find a:

$a^2 + b^2 = c^2$

$a^2 + 6^2 = 9^2$

$a^2 + \quad =$

$a^2 = \quad - \quad =$

$a =$ cm

Now answer these questions on paper. Draw a diagram for each question:

In questions **3–8** find x:

3.

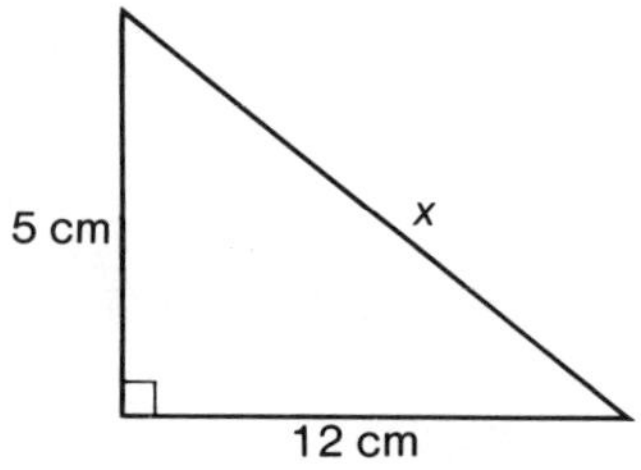

4.

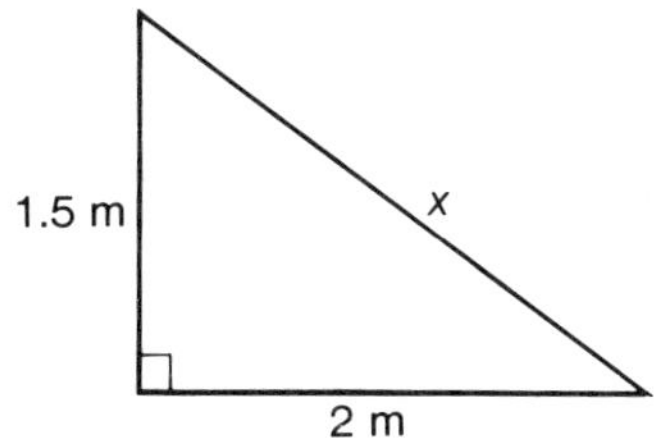

5.

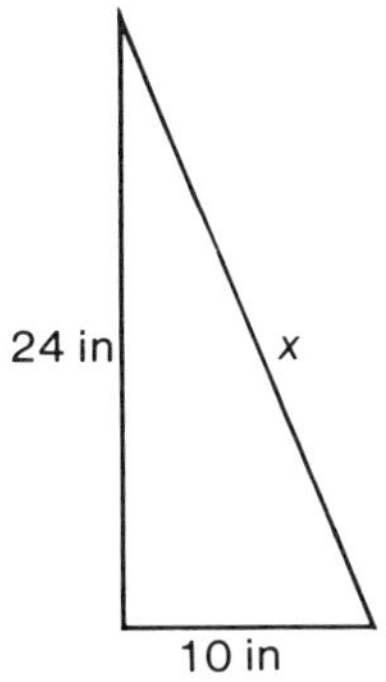

6.

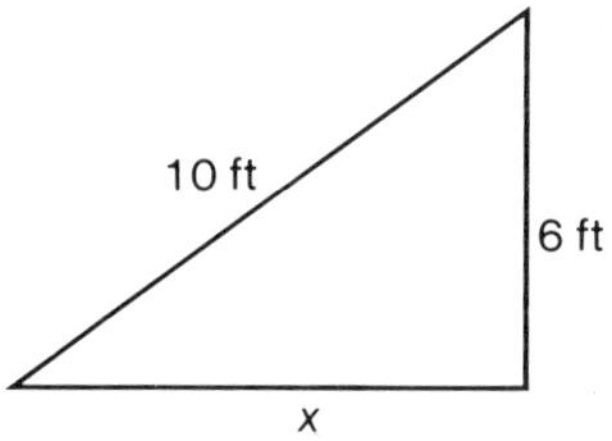

7.

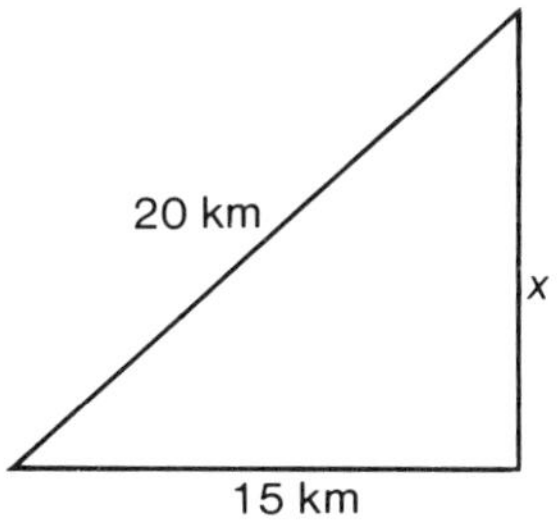

8.

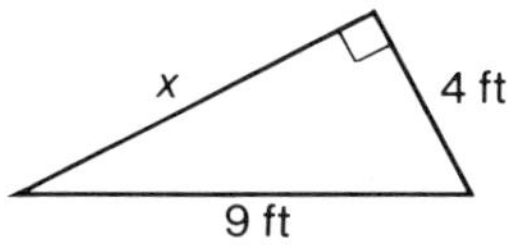

9. A ladder 5 m long is placed against a wall. The lower end is 3 m from the wall. How far up the wall does the ladder reach?

10. A helicopter flies 12 km due North then 5 km due East, and lands. How far is it now from the starting point by a straight route?

B Area of Triangles and Circles

From Unit Ten you should already know how to find the area of a triangle.

$A = \frac{1}{2} \times b \times h$

Can you remember the formulae for a circle?

$A =$

$C =$

Check your answers below.
If you got them wrong *learn* them.
Do these further examples:

1. Find the area of a circle radius 7 cm $\left(\pi = \frac{22}{7}\right)$.
2. A circle has circumference 20 cm. Find its area ($\pi = 3.14$).
3. A baker has 50 round cakes 20 cm in diameter. She wishes to cover the tops of them all with icing. She estimates a 500 g packet of icing sugar will cover 1000 cm^2 of cake.
   How many packets of icing sugar will she need?
4. 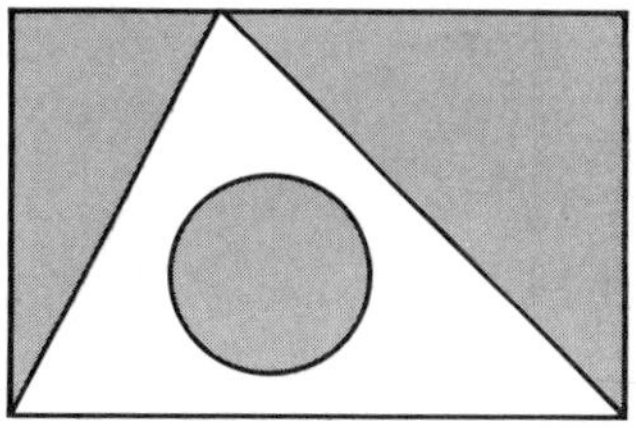

   A carpenter has a piece of wood 2 m wide and 3 m long. He cuts off two triangles as shown then cuts out a circle radius 50 cm. Find the area left (unshaded).

Area of circle $= \pi \times r \times r$
Circumference of circle $= 2 \times \pi \times r$

C Volume

Volume is measured in cubic units i.e. cm^3, m^3, ft^3, etc. A cubic centimetre measures 1 cm in each direction.

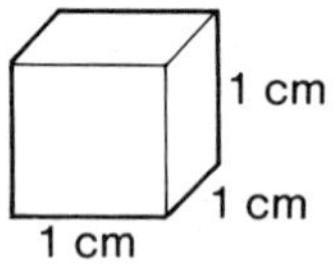

$10\text{ mm} = 1\text{ cm}$

so $10 \times 10 \times 10\text{ mm}^3 = 1\text{ cm}^3$

$1000\text{ mm}^3 = 1\text{ cm}^3$

When calculating volumes:
You *must* have all measurements in the *same* units.
It is best to work in the same units as the answer.

Volume of box (cuboid)

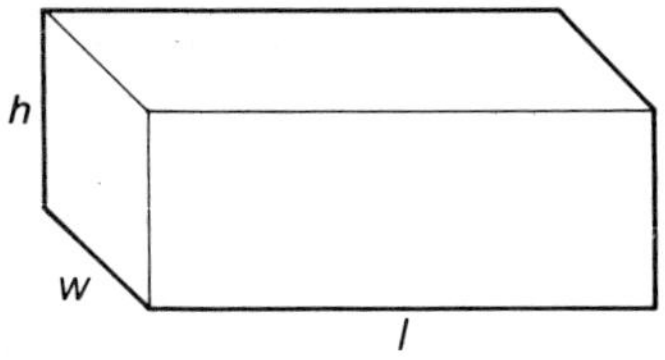

$V = l \times w \times h$

Example

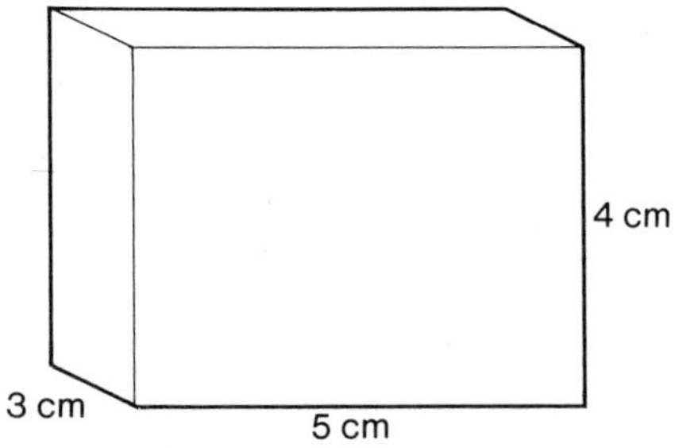

$$\mathbf{V = l \times w \times h}$$
$$= 5 \times 3 \times 4$$
$$= \mathbf{60\ cm^3}$$

The volume of any solid with the same cross-section all the way along is
$V =$ (area of end) $\times$ length
This applies to all the following:

Cylinder

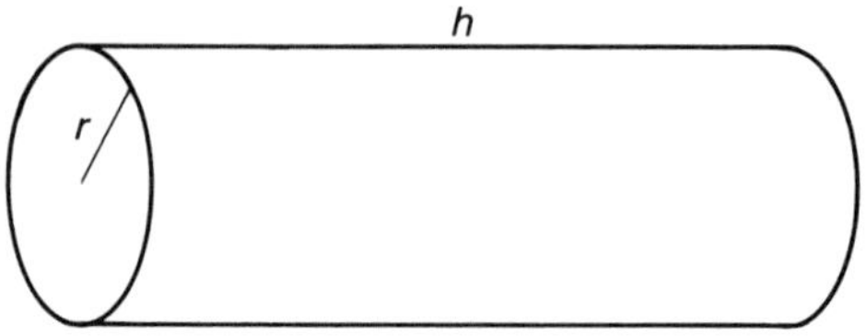

Volume = area of circle $\times$ length

$$\mathbf{V = \pi \times r \times r \times h}$$

Example

Find the volume of a tin diameter 7 cm, height 10 cm ($\pi = 3.14$).

$r = 3.5$ cm

$V = \pi \times r \times r \times h$

$= 3.14 \times 3.5 \times 3.5 \times 10$

$= \mathbf{384.65\ cm^3}$

Triangular Prism

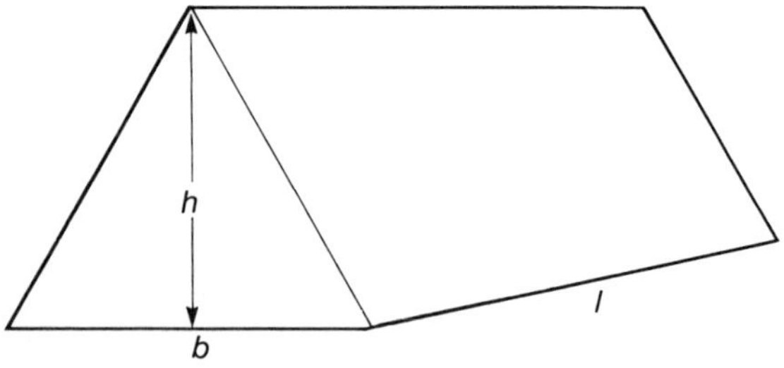

Volume = area of triangle $\times$ length

$$\mathbf{V = \frac{1}{2} \times b \times h \times l}$$

Volume of triangular prism

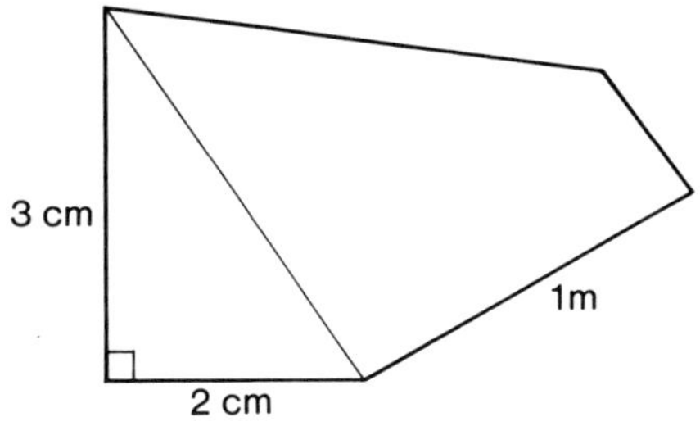

Example

Find the volume of the prism drawn.

1 m = 100 cm

V = *area of end* × *length*

$= \frac{1}{2} \times b \times h \times l$

$= \frac{1}{2} \times 2 \times 3 \times 100$ (*all units in centimetres*)

$= \mathbf{300\ cm^3}$

Fill in these examples:

1. *Find the volume of the dolls' house shown:*

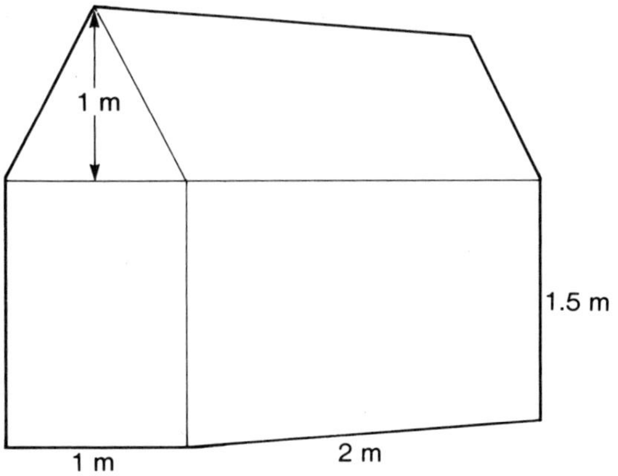

The roof is a triangular prism.

The base is a cuboid.

Volume of cuboid $= l \times w \times h$

$= 2 \times 1 \times 1.5\ m^3$

$=$ m^3

Volume of prism $= \frac{1}{2} \times b \times h \times l$

$= \frac{1}{2} \times 1 \times 1 \times 2\ m^3$

$=$ m^3

Total volume $=$ $+$

$=$ $\mathbf{m^3}$

2. *A stick of rock has a diameter of 1.5 cm. It is 30 cm long. Find its volume* ($\pi = 3.14$):

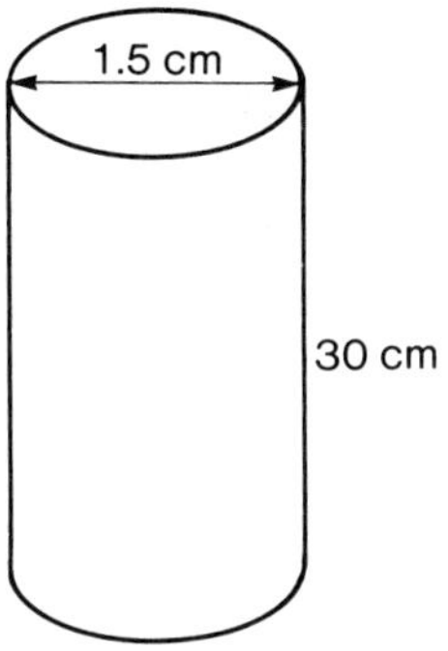

It always helps to draw a diagram.

Radius $= 1.5 \div 2 =$ cm

Volume $= \pi \times r \times r \times h$

$= 3.14 \times$ $\times$ $\times$ cm^3

$=$ $\mathbf{cm^3}$

Now try these on paper:

3. Find the volume of a box 20 cm wide, 50 cm high and 40 cm long.
4. Find the volume of a tin of beans if the diameter is 7 cm and the height 11 cm $\left(\pi = \frac{22}{7}\right)$.
5. Find the volume of this tent.

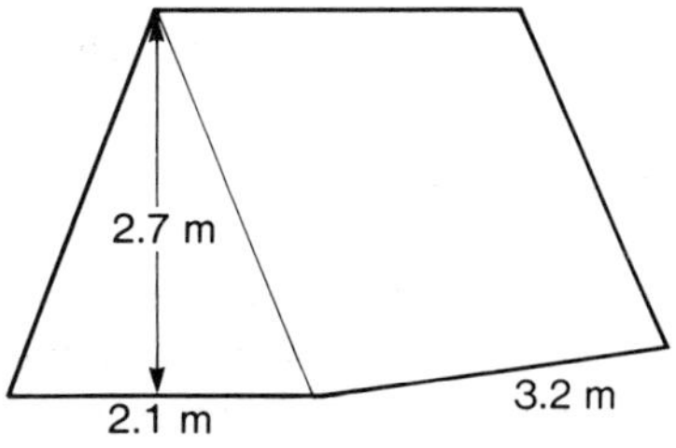

6. A rectangular oil tank is 2.3 m long, 1.8 m wide, and is filled to a depth of 80 cm with oil. Find the volume of oil in m^3.
   The next week the oil is only 45 cm deep. How much oil does it now contain?
7. Find some household objects with a uniform cross section (e.g. tins, boxes). Draw them, estimate or measure their size and calculate their volume.

D Surface Area

The surface area is the total area of all the faces. Find each area in turn and add them up. It helps to draw a diagram.

Example

Find the surface area of the shape drawn:

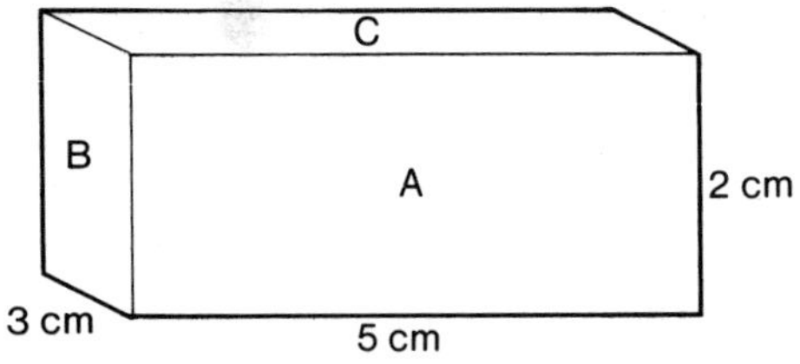

Area A $= 5 \times 2 = 10$ cm²

Area B $= 3 \times 2 = 6$ cm²

Area C $= 3 \times 5 = 15$ cm²

31 cm²

There are two faces of each size.

Total area $= 31 \times 2 =$ **62 cm²**

Fill in this example:

1. *A room with the dimensions shown is to be painted on the walls and ceiling. Assuming the door and window cover a total of 36 ft² find the area to be painted.*

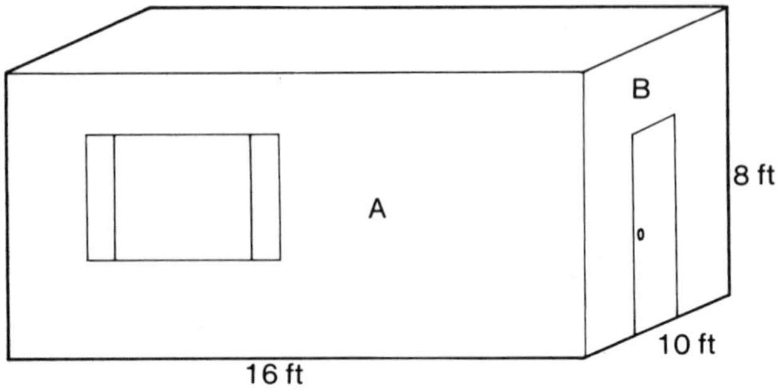

Area of ceiling $= 10 \times 16 =$ ft²

Area of wall A $16 \times 8 =$ ft²

Area of wall B $=$

Total area $=$ $+ 2 \times$ $+ 2 \times$ $- 36$

$=$ ft²

If a litre of paint covers 250 ft² how much paint is required for two coats if it is bought in 1 litre tins?

Total area to be painted in 2 coats $= 2 \times$ ft²

Paint needed $= \frac{\quad}{250}$ ℓ

$=$ ℓ (*whole number of litres*).

Now try these on paper:

2. A garden shed rests against the wall of a house. The three sides and roof and door are to be painted. Find the area to be painted.

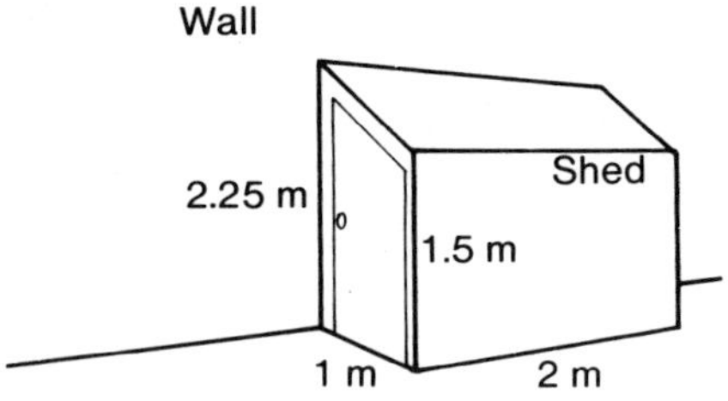

3. Find the surface area of a cube of side 4 cm.
4. Find the total surface area of a box dimensions 1.8 cm, 3.9 cm, 6.1 cm.
5. Find the surface area of the dolls' house in section C, including the floor. (Use Pythagoras' theorem to find the sloping edge.)

E Capacity

Capacity is the amount a container will hold. It is sometimes measured in the same units as volume i.e. cm^3, m^3, but often litres (metric) or pints (imperial) are used.

$1\ \ell = 1000\ cm^3$

$1000\ \ell = 1\ m^3$

$1000\ m\ell = 1\ \ell$

Example

A tank measures 50 cm by $1\frac{1}{2}$ m by 2 m.

How many litres will it hold?

Work in centimetres.

$$\begin{aligned} \textit{Volume} &= 50 \times 150 \times 200 \\ &= 1500\,000\ cm^3 \\ &= \mathbf{1500\ \ell} \end{aligned}$$

Try these questions yourself:

1. A bottle contains $1500\ cm^3$. Express this in litres.
2. Find the capacity of an oil drum, diameter 70 cm, height 1.2 m $\left(\pi = \frac{22}{7}\right)$.
3. Find the amount of water in a rectangular swimming pool 10 m by 8 m and 0.6 m deep:
   a) in litres
   b) in gallons (1 gallon = 4.5 litres).
4. An oil tank measures 1.32 m long, 1.22 m wide by 1.76 m high. The depth of oil in the tank is 23 cm.
   The tanker delivers oil in 100 ℓ quantities. What is the most oil the customer can order without overflowing the tank?
5. A man wants to concrete his drive to a depth of 6 cm. The drive measures 3.8 m by 13 m. How much ready mix concrete must he order. (Answer in cubic metres.)

UNIT EIGHTEEN **Test**

Set your work out neatly on lined paper. Put your name at the top.

1. A window box has external measurements 120 cm long, 20 cm wide and 15 cm high. The wood is 30 mm thick. (It has no lid.)
   a) Calculate the outside surface area of the four sides and base.
   b) Calculate the inside surface area.
   c) Calculate the volume of soil needed to fill in. (Answer in cubic centimetres.)
   d) Find the cost of filling it with compost costing 28 p per litre.
2. Mr. Smith has a rectangular oil tank for his central heating. It is 2.2 m long, 70 cm wide and 1.8 m high.
   a) How many litres does the tank hold?
   b) How many gallons does it hold? (1 litre = 1.75 pints.)
   c) How long will it take to fill the tank if the tanker delivers 150 ℓ per minute? (Answer to nearest minute.)
3. A cube of metal of side 50 cm is melted down to make toy soldiers. Each soldier needs 15 cℓ of metal. 8% of the metal is wasted. How many soldiers are made?
4.

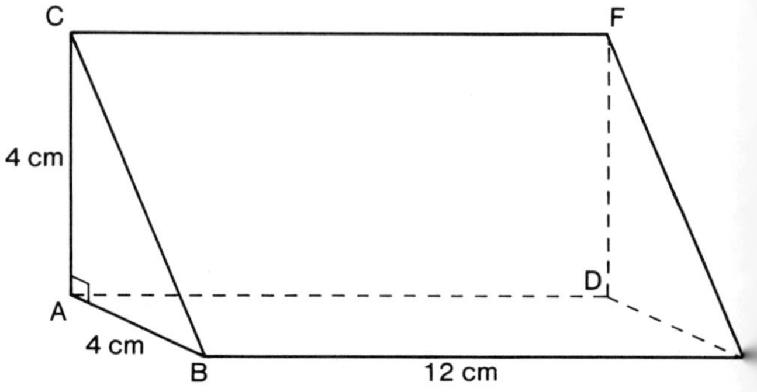

   Calculate:
   a) BC
   b) The volume of the prism
   c) The total surface area.

5. A paper cup is approximately a cylinder radius 2 cm height 9 cm. It is to be 90% full of coke.
   a) Find the volume of the cup ($\pi = 3.14$).
   b) Find the amount of coke in one cup. (Answer in millilitres.)
   c) How many cups of coke can you get from a two-litre bottle?

Do not forget to check your answers.

UNIT EIGHTEEN ANSWERS

A

AC = 3 cm
BC = 4 cm
AB = 5 cm
AC^2 = 9 cm
BC^2 = 16 cm
AB^2 = 25 cm
$AC^2 + BC^2$ = 25 cm

1. $c = 10$ cm
2. $a = \sqrt{45} = 6.7$ cm
3. 13 cm
4. 2.5 m
5. 26 in
6. 8 ft
7. 13.23 km
8. 8.06 ft
9. 4 m
10. 13 km

B

1. 154 cm^2
2. 31.85 cm^2
3. Area of one cake = 314 cm^2, area of 50 cakes = 15 700 cm^2 = 15.7 packets = **16** packets
4. Area = 3 m^2 − .785 m^2 = **2.215 m^2**

C

1. Volume of cuboid = 3 m^3, Volume of prism = 1 m^3, Total = 4 m^3
2. $r = 0.75$ cm, $V = 53.0$ cm^3
3. 40 000 cm^3 or 0.04 m^3
4. 423 cm^3
5. 9.07 m^3
6. 3.312 m^3, 1.863 m^3

D

1. Area of ceiling 160 ft^3, area of wall A = 128 ft^2, area of wall B = 80 ft^2, total area = 540 ft^2, paint needed = 5 ℓ
2. 9.25 m^2
3. 96 cm^2
4. 83.58 cm^2
5. 16.5 m^2

E

1. 1.5 ℓ
2. 462
3. a) 48 000 ℓ b) 10 667 gal
4. 2400 ℓ
5. 3 m^3

UNIT NINETEEN

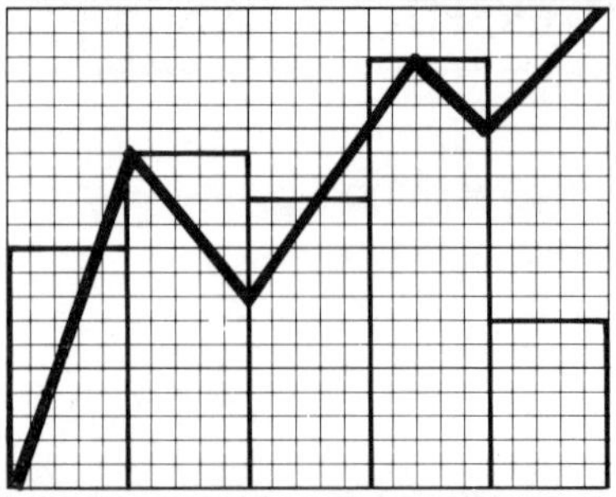

Graphs and Charts

Graphs and charts present information in a form more easily understood than words. You need to know how to draw graphs and charts and how to use them.

A Bar Charts

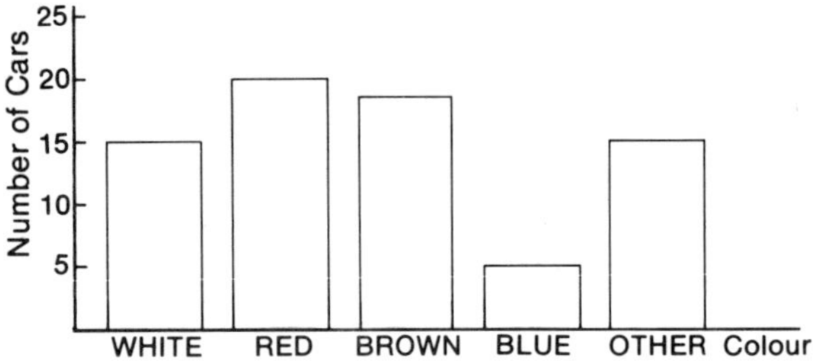

Bar-chart to show the colour of cars passing a school between 9 a.m. and 9.30 a.m.

Rules for drawing a bar chart:

Label the axes.
Make all the bars the same width.
The height of the bar corresponds to the frequency.
Bar charts are sometimes drawn horizontally e.g.

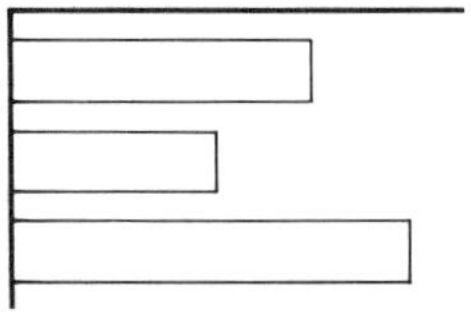

When you are reading a bar chart look at the scale carefully.
Gaps between the bars are optional.

Example

Answer the questions referring to the bar chart:

1. *How many blue cars passed the school?*
2. *What was the total number of cars passing the school?*
3. *What percentage of cars were white?*
4. *How many more brown cars were there than blue ones?*
5. *What was the average number of cars passing the school per minute?*

Check your answers.
Try this yourself:
Draw a bar chart to show the sunshine figures for "Sunnydays" Holiday Village.

Month	A	M	J	J	A	S	O
Average hours per day	4	6	8	9	$7\frac{1}{2}$	6	5

Draw a line to show the average hours of sunshine for the whole period.

B Pie-charts

How to draw a pie-chart:

1. Draw a large circle.
2. Calculate the angles (check they add up to 360°). The example shows you how.
3. Measure the angles carefully with a protractor.
4. Label the sectors.

Example

30 students buy a drink from a vending machine.

14 buy coffee

6 buy tea

5 buy soup

5 buy chocolate

The circle has 360° to represent 30 students.
To find the angles multiply each number of drinks by $\frac{360}{30}$

$$coffee = 14 \times \frac{\overset{12}{\cancel{360}}}{\cancel{30}} = 168°$$

$$tea = 6 \times \frac{\overset{12}{\cancel{360}}}{\cancel{30}} = 72°$$

$$soup = 5 \times \frac{\overset{12}{\cancel{360}}}{\cancel{30}} = 60°$$

$$chocolate = 5 \times \frac{\overset{12}{\cancel{360}}}{\cancel{30}} = 60°$$

Total = 360° (Check)

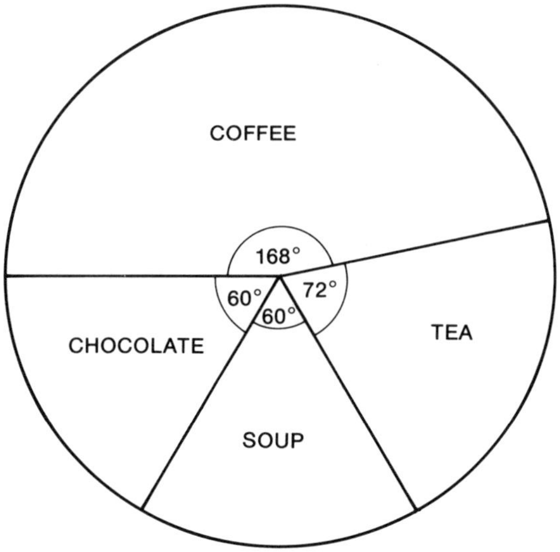

Pie-chart to show the drinks purchased from a machine.

Always show the angles you have calculated either marked on your pie-chart or in your working.

Try these yourself:

1. Draw a pie-chart to show the results of this newspaper survey:

Times	5
Telegraph	15
Mirror	22
Sun	28
Express	2

2. Often you are given a pie-chart to measure the angles and answer questions. Measure the angles carefully and make sure your figures add up to 360° then answer the questions.

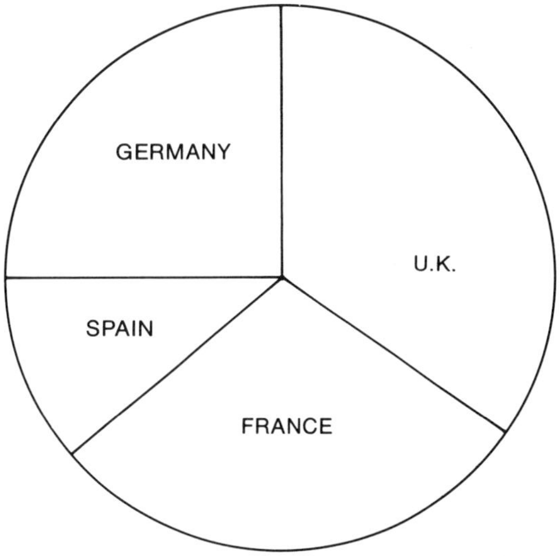

A company sells its goods in UK, Germany, Spain and France. In 1983 it sold £72 000 worth of goods.

a) How many pounds worth of goods were sold in each country?
b) Which country bought most goods?
c) What percentage were sold in France?
d) How much more was sold in UK than Spain?
e) If each item sold for £5, how many were sold in Germany?

C Pictograms

Example

A company employs 300 people. There were 150 men, 70 women, 20 boys and 60 girls.

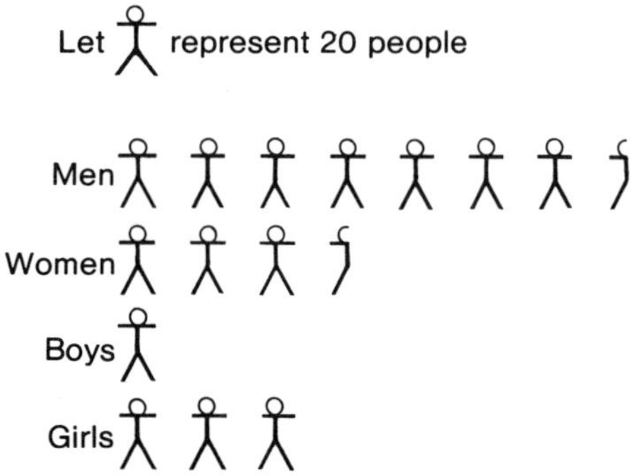

Now draw a pictogram to represent this information: The number of beefburgers sold by MacWhimpey on each day of the week is as follows:

M	50
T	150
W	300
Th	220
F	400
S	550
S	350

D Graphs

If you are given a scale use it. Label the axes. Plot the points with crosses and join them.

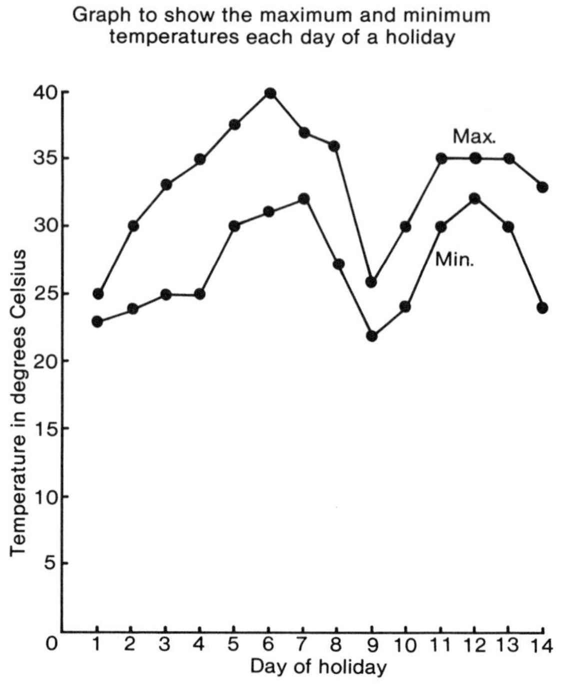

Answer these questions:

1. On which day was the hottest temperature reached?
2. On which day was the coldest temperature reached?
3. Which day had the biggest range of temperatures?
4. Plot on the graph the average temperature for each day.
5. On which days was it hotter than 35 °C?
6. Which days stayed colder than 30 °C all day?

UNIT NINETEEN **Test**

Set your work out neatly on lined paper. Put your name at the top.

1. During the sales a shop sold goods to the following values in each department:

Ladieswear	£15 000
Menswear	£8 000
Childrenswear	£10 000
Household	£3000

   Represent this on a pie-chart.

2. A survey of children's pets showed that a class of children owned between them: 6 ponies, 15 small rodents, 20 fish, 8 dogs and 23 cats. Represent this on a pie-chart.

3.

Pictogram to show number of new houses built

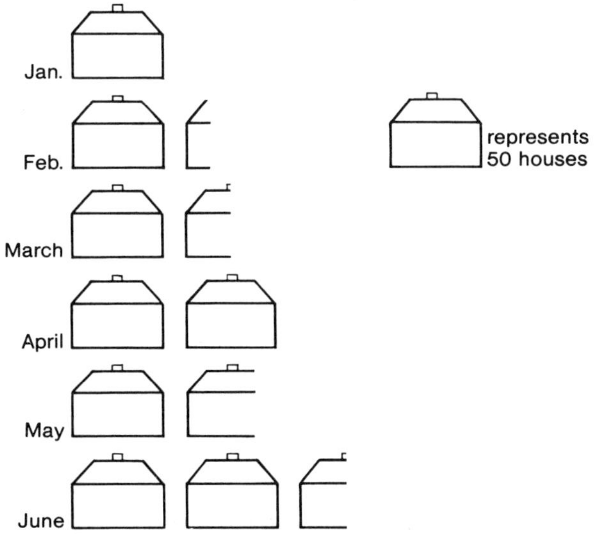

   Estimate the number of houses built in each month from January to June.

4. The number of cars produced by a factory each week is as follows:

Week Number	1	2	3	4	5	6	7	8
Cars	25	17	36	22	34	5	28	17

   Represent the information on a graph. Answer these questions:

   a) During which week was production lowest?
   b) Find the average number of cars produced per week.
   c) By what percentage did the number of cars produced in week three exceed the average?

Do not forget to check your work.

A

1. 5 2. 75

3. $\frac{\overset{5}{\cancel{15}}}{\underset{3}{\cancel{75}}} \times \frac{\overset{4}{\cancel{100}}}{1} = 20\%$

4. 13

5. $\frac{75}{30} = 2.5$ cars per minute

Average sunshine $= \frac{45.5}{7} =$ **6.5 hours**

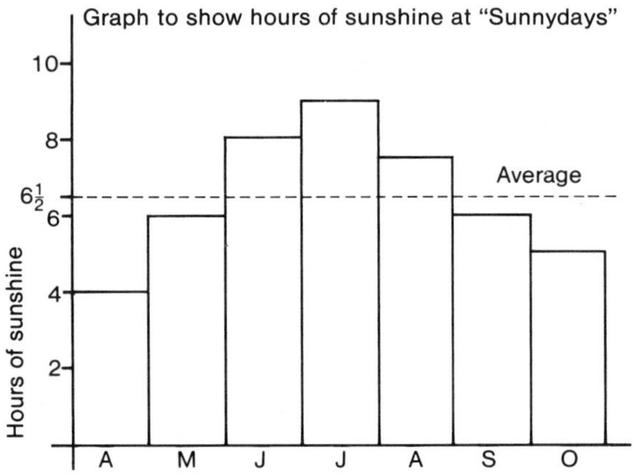

B

1. 25°, 75°, 110°, 140°, 10°

2. a) UK £25 000, Germany £18 000
Spain £8000, France £21 000
b) UK c) 29.2%
d) £17 000 e) 3600

C

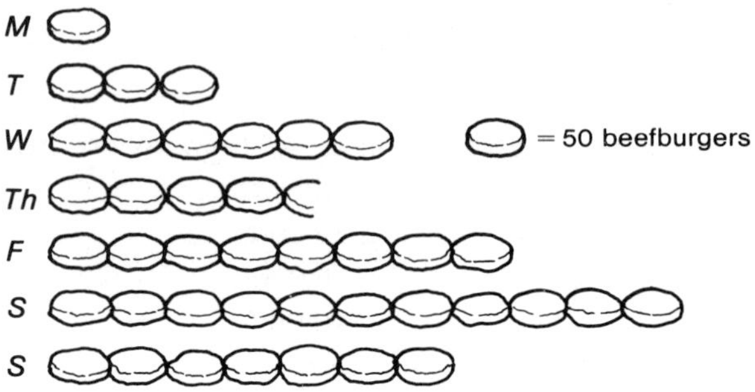

D

1. 6 2. 9 3. 4
5. 5, 6, 7, 8 6. 1, 9

Graph to show the maximum and minimum temperatures each day of a holiday

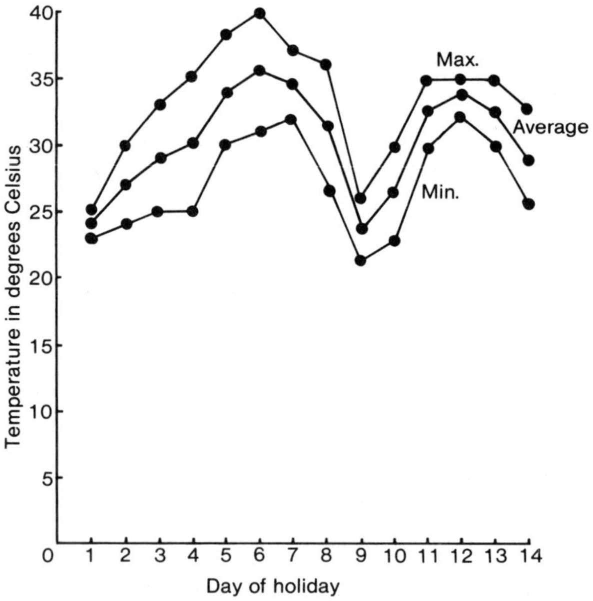

UNIT TWENTY

Revision

You should now practise past examination papers. The papers in this Unit are similar to questions in examination papers.
You have already done Paper I questions in Unit Thirteen. Now do Paper II.
Read these pages *carefully* first.

The Exam Paper

Always read the front of the paper carefully at the beginning of the exam. The instructions are not always the same. Look for the following:

a) How long is the exam?
b) How many questions do I have to answer?
c) Do I get marks for working?
d) Can I use a calculator?

Get a recent Paper II and *write down* the answers to the above questions.
You are given ten minutes reading time before the examination starts. Use the time carefully.

a) Read the questions carefully.
b) Decide which ones you are able to answer.
c) Decide the order you are going to answer them.
d) Think about how to do your first question.

The Questions

Each question has several parts, tackle one part at a time. Show your working clearly and *check* each calculation you make. When you think you have finished read the *question* again. Have you answered it? Does it say, for instance, give the answer to the nearest pound, or find the new price when you have found the extra cost.
Don't waste time on a part you cannot do: go on to another question. You may have time to come back to it.

Example

Here is an example of an exam type question with part of the answer filled in for you. It will show you how much working you should give.

A girl decides to buy a music centre. She goes to three shops.

	Cash Price	*HP Price (Monthly payments)*
Shop A	£240.00	£21.50 deposit, 12 payments of £19.90
Shop B	£249.99	one third deposit, 24 payments of £7.29
Shop C	£259.80	25% deposit, 18 payments of £11.50

a) *Calculate the total HP price at each shop.*
b) *How much more is each HP price than the cheapest cash price?*
Express this extra cost as a percentage of the cheapest cash price. (Answers to 2 decimal places.)
c) *If the manufacturer sells the music centre for £160.00 which shop makes the most profit when selling three music centres for cash and one on HP? Express the profit as a percentage of the cost price. (Answer to 1 decimal place.)*

a) *Shop A* Deposit £21.50
Repayments £19.90 × 12 = £
Total cost £21.50 + =

Shop B Deposit £249.99 ÷ 3 =
Repayments £7.29 × 24 =
Total cost =

Shop C Deposit $\frac{25}{100}$ × £259.80 =
Repayments £11.50 × 18 =
Total cost = ANS

b) The cheapest cash price is £240.00
Shop A − £240.00 =

Percentage = $\frac{\quad}{£240.00} \times \frac{100}{1}$
=
= (2 decimal places) ANS

Shop B − £240.00 =

Percentage $\frac{\quad}{£240} \times \frac{100}{1}$
= (2 decimal places) ANS

Shop C − £240 =

Percentage = $\frac{\quad}{£240} \times \frac{100}{1}$
= (2 decimal places) ANS

c) Shop C is the dearest.
Cash 3 × £ =
HP 1 × =
Total selling price = £
Cost Price = 4 × £160 = £
Profit = − =
Percentage profit = $\frac{\text{Profit}}{\text{Cost price}} \times \frac{100}{1}$
=
= (1 decimal place) ANS

Now work through the Paper II you have chosen. See how much you can do in the time allowed and then carry on and do as much as you can.
Mark your work and correct any questions you got wrong.
Ask for help with any questions you cannot do.
As you get used to the type of questions set in examinations you will get better at answering them. Lots of practice is the best way to pass the examination.

Good luck.

ANSWERS TO EXAMPLE

a) A. £260.30 B. £258.29 C. £271.95
b) £20.30 = 8.46%; £18.29 = 7.62%; £31.95 = 13.31%
c) £1051.35; £640; £411.35; 64.3%

Paper Seven

Time allowed 1 hour.
You may use a calculator but show your methods clearly.
Answer 5 questions only.

1. A girl wants to buy a Musico stereo system. In a shop it costs £399 cash or £125 deposit and 12 monthly payments of £26.99. If she pays cash she can borrow the money from her bank at 10.5% interest over one year.
   a) How much will she pay altogether if she chooses the shop's HP plan?
   b) How much will she pay if she borrows from the bank?
   c) Which is the cheapest and by how much?
2. a) The rateable value of a house is £415 and the rates are 115 p in the pound. How much are the rates per week?
   b) A town has a rateable value of £269 000. It is necessary to increase the rates by 2 p in the pound. How much extra income will this bring in?
3. A man is paid £102.90 for a 39 hour week which includes 3 hours overtime. If overtime is paid at double the normal rate then:
   a) What is the normal rate per hour?
   b) If he earns £124.95 in the next week then how many hours has he worked?
4. A holiday costs £279 per adult. Children aged five to eleven get a 20% reduction. Children under five get a 60% reduction.
   A man goes on holiday with his wife and two children aged six and four. How much does he pay altogether?
   When he books his holiday he pays 10% deposit and £9.50 insurance per person (including children). How much does he pay when booking?
5.

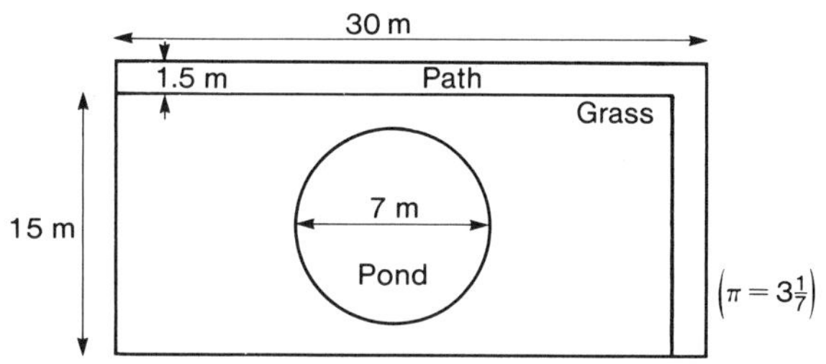

   A garden is 30 m by 15 m. The path is 1.5 m wide. The pond is 7 m in diameter.
   a) Find the area of grass $\left(\pi = \frac{22}{7}\right)$.
   b) Find the total cost of the path at £8.50 per square metre.
6. During the day a child spends 2 hours eating, $3\frac{1}{2}$ hours playing at home, $1\frac{1}{2}$ hours going for a walk, 3 hours at playgroup and the rest asleep. Draw a pie-chart to show this information. State clearly the angle of each sector.
7. A household uses 315 therms of gas in a quarter. The gas board make a £9 standing charge and 22.4 p per therm.
   a) Calculate the total bill.
   b) The next quarter the standing charge goes up to £9.50 and the therms to 24.2 p. In order not to increase the total bill what is the maximum number of therms that can be used?

Paper Eight

Time allowed 1 hour.
Answer 4 questions only.

You may use a calculator but show all necessary working.

1. A car ferry company offer the following rates:

Return Fare (if your journey covers two periods take the average cost)	Oct 1–28	Oct 29–Dec 31	Jan 1–Feb 28
Cars up to 4 m long	£46.50	£41.00	£28.00
Cars 4 m to 6 m long	£61.50	£49.00	£28.00
Each extra metre or part thereof	£10.00	£10.00	£10.00
Adults	£14.00	£13.00	£14.00
Children	£9.00	£8.00	£9.00

a) Calculate the return fare for Mr. and Mrs. Smith and a 5 m car going on Oct 7th and returning on Oct 15th.
b) Calculate the return fare for the Jones family of two adults and two children with a 5.5 m car going on Dec 28th and returning on Jan 7th.
c) How much would the Jones family save if they changed their holiday dates to Jan 5 until Jan 14th? Express this saving as a percentage of the original fare.

2. A finance company will loan £2000 with the following choice of repayments:
a) £74.13 per month for 36 months.
b) £52.77 per month for 60 months.
c) £43.39 per month for 90 months.
Calculate the total paid in each case and the equivalent rate of Simple Interest.
If instead I borrow £2000 from the bank I have to pay 15% Compound Interest after 3 years. How much interest would I pay the bank?

3. A man buys a car for £4900. He pays £148 to tax and insure it per year. He drives 100 miles per week. The car averages 28 miles per gallon. Petrol costs £1.86 per gallon. His repair bills cost £12 per month. He sells the car after 2 years for £2630.
a) Find the total petrol costs for the two years.
b) Find the total cost of owning the car for two years.
c) If instead of buying a car he spends £1.46 per day five days a week on bus fares and £5 each weekend on a taxi, how much less would he spend per week? (1 year = 52 weeks.)

4. Mr. Dig buys a new house with a garden, and he plans the garden as shown in the diagram. The garden is covered with grass except for 1 m round the edge which is either path or roses.

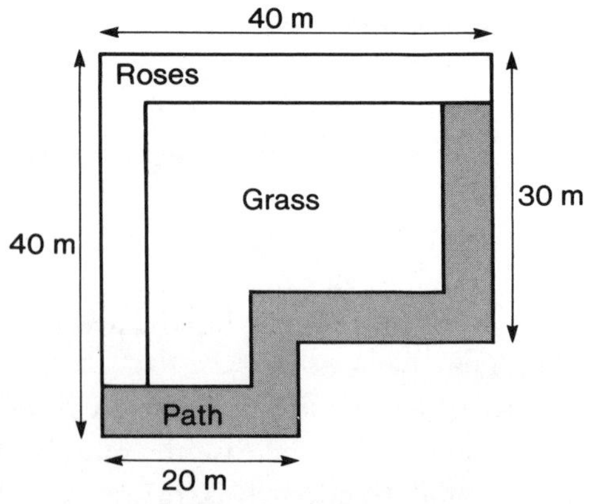

a) Find the area of the grass.

b) Find the cost of the path at £2.50 per square metre.

c) He plants one rose every 2 m^2. How many roses does he buy?

d) Turf costs 49 p per metre and roses cost £1.99 each. Find the total cost of roses, grass and path.

5.

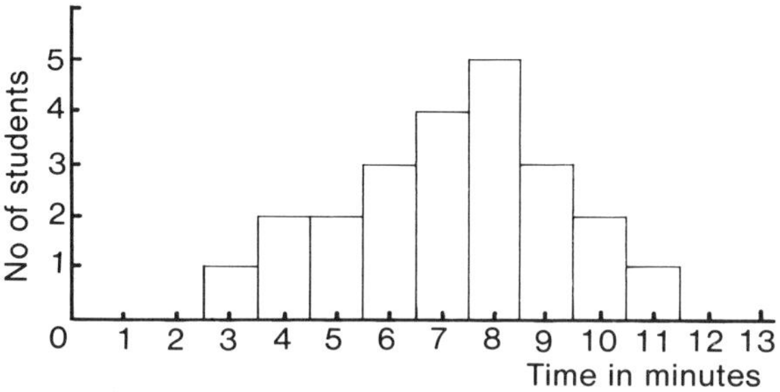

The bar chart shows the times taken by a group of students to run 1000 m.

a) How many students took part in the run?

b) Find the average time taken.

c) What percentage of students ran the 1000 m in under $6\frac{1}{2}$ minutes?

6. A motorist leaves London at 8.40 a.m. and travels for 20 miles until he reaches the motorway. Then he travels 320 miles on the motorway. He averages 70 m.p.h. on the motorway and 15 m.p.h on other roads. He makes two stops of 20 minutes each.

a) At what time does he arrive?

b) He leaves to come home at 1500 next day and arrives home at 2250. Find his average speed on the journey home.

c) His company pay him 22 p per mile expenses and £12 for all his meals on the way. How much does he claim from the company altogether?

Paper Nine

Time allowed $1\frac{1}{2}$ hours.
Calculators may be used but candidates must show any numerical expression being evaluated by calculator or marks will be lost.
Answer any 5 questions.

1. a) A man invests £100 at $12\frac{1}{2}\%$ per year Compound Interest. How much will it amount to after (i) 1 year, (ii) 2 years?
   b) How much will he have to invest to amount to £2000 after 2 years? (Answer to nearest ten pounds.)
   c) He spends the £2000 as a deposit on a car costing £5850. He borrows the rest from the bank at 15% Simple Interest per year over the whole loan period and pays it back in 36 monthly payments. Find
      (i) The total interest paid to the bank.
      (ii) The cost of each monthly payment. (Answer to the nearest penny.)

2. A pointer is cut out of sheet metal as in the diagram. The metal sheet is 14 cm by 27 cm.

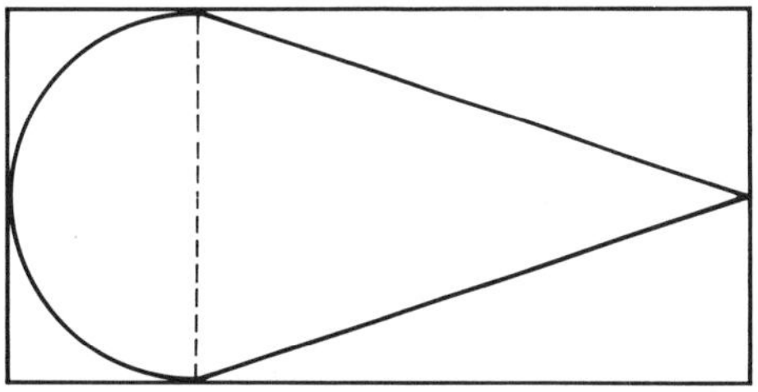

Find
   a) The area of the semicircle $\left(\pi = \frac{22}{7}\right)$.
   b) The area remaining after the shape is cut out.
   c) The perimeter of the shape.

3. A small hotel employs two managers, a chef, six cleaners and four waiters. The managers and the chef are each paid a salary of £12 000 a year. They are paid in 52 weekly parts. The cleaners are paid £1.52 per hour and work 22 hours each per week. The waiters are paid £1.90 per hour for the first 30 hours and time and a half for

overtime. The waiters average 36 hours per week.

a) Find the hotel's annual wage bill. (Answer to the nearest hundred pounds.)

b) During a busy week a waiter works 51 hours. How much will he be paid for that week?

c) During a pay increase everyone gets an 8% rise. How much is the hotel's average weekly wage bill now?

4. Assume exchange rates are as follows and work each calculation to the nearest penny.
France £1 = 11.35 francs
Germany £1 = 4.15 marks
The bank charges 1% on each transaction.

a) An American arrives in London with £220, changes it all into francs and goes to France. How many francs does he receive?

b) He spends 72 francs in France and returns to England where he changes the remainder of his money to pounds. He spends some money in England and has £105 left. How much does he spend in England?

c) He changes the remaining money into marks and goes to Germany where he spends 300 marks. He converts the rest back to pounds. How much is left?

5. A cyclist wishes to travel from Pietown to Queuetown, a distance of 40 miles. He sets off at 8 a.m. and cycles for 1 hour at 12 m.p.h. He rests for 15 minutes then travels a further 15 miles in $1\frac{1}{2}$ hours. He now goes into a café for half an hour. After this break he completes the journey arriving at Queuetown at 12.15 p.m. Draw a graph to represent his journey.
Calculate his speed for the final stage of the journey.
A car left Pietown at 10 a.m. and travelled to Queuetown at a constant speed. He passed the cyclist just as he was starting his break at the café. Draw a line to represent the motorist's journey and find what time he arrived at Queuetown.

6. A Building Society shows in its Annual Report that the income for 1984 was £596 000 000. This came from Mortgages, Investments and other sources in the ratio 9 : 2 : 1. Calculate how much came from each source to the nearest million pounds and represent the information on a pie-chart.

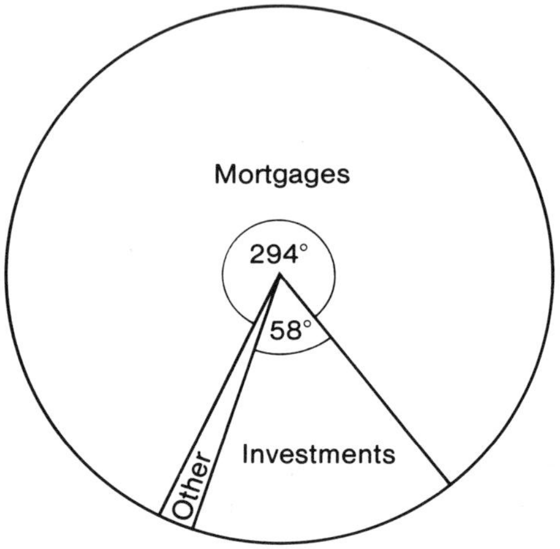

This pie-chart shows the society's assets. If mortgages total £4602 000 000 calculate the investments and other assets. (Answer to nearest million pounds.)

7. Three banks offer the following special offers to young savers:

a) Bank A gives a free bag and books worth £15 and 4% interest.

b) Bank B gives a £5 bonus after one year and 8% interest.

c) Bank C gives no gifts but offers 15% interest.

A boy has £20 to invest. How much value would he get in cash or goods from each bank after one year. Which is the best bank to choose?

Extra Examples

You can do these extra examples as you work through the course, or at the end. When you are working through the course, if you need more practice on any topic choose the corresponding set of extra examples.
After you have completed the course do the examples belonging to the Units that you wish to revise. Look back at the earlier work as necessary.
Some of these questions are taken from past RSA papers. They are marked RSA and are to be found at the end of each set of examples.

UNIT ONE

A

1. 15 + 12
2. 59 + 16
3. 82 + 19
4. 105 + 96
5. 214 + 36
6. 511 + 215
7. 836 + 99
8. 67 + 213 + 223
9. 914 + 8 + 59
10. 96 + 59 + 137

B

1. 35 − 12
2. 98 − 54
3. 214 − 82
4. 98 − 79
5. 197 − 38
6. 863 − 76
7. 342 − 178
8. 456 − 98
9. 1010 − 872
10. 5000 − 2479

C

1. 8 × 9
2. 9 × 2
3. 7 × 3
4. 8 × 7
5. 9 × 5
6. 9 × 9
7. 8 × 6
8. 5 × 4
9. 4 × 7
10. 9 × 2

D

1. 36 ÷ 6
2. 16 ÷ 4
3. 18 ÷ 2
4. 27 ÷ 3
5. 36 ÷ 9
6. 81 ÷ 9
7. 72 ÷ 8
8. 30 ÷ 5
9. 48 ÷ 6
10. 56 ÷ 7

E

1. 72 × 2
2. 81 × 3
3. 25 × 4
4. 16 × 5
5. 114 × 6
6. 215 × 7
7. 95 × 8
8. 37 × 9
9. 115 × 8
10. 297 × 6

F

1. 9 × 10
2. 4 × 10
3. 10 × 100
4. 92 × 10
5. 100 × 84
6. 89 × 1000
7. 100 × 14
8. 1000 × 62
9. 416 × 10
10. 518 × 1000

G

1. 15 × 20
2. 19 × 40
3. 86 × 200
4. 17 × 12
5. 29 × 82
6. 94 × 34
7. 146 × 37
8. 249 × 68
9. 128 × 360
10. 721 × 114

H

1. 880 ÷ 4
2. 2450 ÷ 5
3. 582 ÷ 3
4. 2304 ÷ 9
5. 1962 ÷ 2
6. 7616 ÷ 8
7. 3647 ÷ 7
8. 3588 ÷ 4
9. 5886 ÷ 9
10. 3948 ÷ 6

I

1. 20 ÷ 10
2. 100 ÷ 10
3. 550 ÷ 10
4. 2000 ÷ 10
5. 3400 ÷ 100
6. 8000 ÷ 100
7. 276 000 ÷ 100
8. 51 000 ÷ 1000
9. 87 600 ÷ 10
10. 310 000 ÷ 1000

J

1. 195 ÷ 13
2. 325 ÷ 25
3. 1728 ÷ 27
4. 1344 ÷ 24
5. 1575 ÷ 21
6. 3648 ÷ 38
7. 1938 ÷ 17
8. 5635 ÷ 23
9. 17 272 ÷ 68
10. 98 901 ÷ 99

K Complete the series:

1. 10, 11, 12, 13, __, __
2. 25, 24, 23, 22, __, __
3. 24, 26, 28, 30, __, __
4. 10, 12, 14, 16, __, __
5. 5, 7, 9, 11, __, __
6. 12, 13, 15, 18, __, __
7. 18, 17, 15, 12, __, __
8. 16, 24, 32, 40, __, __
9. 49, 42, 35, 28, __, __
10. 1, 3, 7, 15, __, __

L Mixed Examples:

1. 93 + 715
2. 30 × 3
3. 814 − 93
4. 69 × 18
5. 72 ÷ 8
6. 176 ÷ 4
7. 104 ÷ 13
8. 718 × 1000
9. 514 × 21
10. 914 + 7 + 58
11. 500 − 289 (RSA)
12. Write down the next two numbers in the series 20, 21, 23, 26, 30 (RSA)
13. $\frac{264 + 176}{22}$ (RSA)
14. 12 × 105 (RSA)
15. 27 + 15 + 8 + 139 (RSA)
16. 76 × 17 (RSA)
17. 136 ÷ 17 (RSA)
18. 7304 − 6096 (RSA)
19. 64 × 14 (RSA)
20. 40 + 17 + 6 + 109 (RSA)

UNIT TWO

A

1. 3.6 + 9.2
2. 15.4 − 3.1
3. 18.4 + 8.67
4. 9.7 − 3.8
5. 11.48 + 6.7
6. 12.6 − 3.71
7. 25.1 + 7.2 + 12.5
8. 31.6 − 17.8
9. 200 + 6.72
10. 200 − 6.72

B

1. 5.8 × 10
2. 83.6 × 100
3. 21 × 100
4. 8.9 × 100
5. 6.71 × 1000
6. 0.02 × 100
7. 0.0361 × 10
8. 35.92 × 1000
9. 172.1 × 100
10. 101.9 × 100

C

1. 9.8 × 5
2. 3.7 × 2
3. 0.3 × 0.4
4. 17.1 × 1.2
5. 8.7 × 8.3
6. 11.2 × 0.31
7. 15.9 × 1.7
8. 12.34 × 0.021
9. 17.61 × 32
10. 290 × 6.7

D

1. 3.79 ÷ 10
2. 16.38 ÷ 100
3. 0.8 ÷ 10
4. 127.4 ÷ 1000
5. 15.72 ÷ 100
6. 21.7 ÷ 10
7. 15.9 ÷ 10
8. 27.92 ÷ 1000
9. 2.789 ÷ 10
10. 28 ÷ 100

E

Divide exactly:

1. 3.6 ÷ 0.4
2. 2.8 ÷ 0.5
3. 7.62 ÷ 3
4. 8.712 ÷ 0.09
5. 7.711 ÷ 1.1

Work to 3 decimal places:

6. 8.71 ÷ 3.4
7. 7.28 ÷ 0.14
8. 126 ÷ 5.1
9. 63 ÷ 8.4
10. 12.48 ÷ 0.41

F Work to 3 decimal places:

1. 11 ÷ 5
2. 7 ÷ 2
3. 15 ÷ 9
4. 8 ÷ 3
5. 21 ÷ 7
6. 15 ÷ 6
7. 21 ÷ 4
8. 19 ÷ 8
9. 3 ÷ 8
10. 23 ÷ 7

G Mixed Examples:

1. 1.21 + 0.73
2. 3.6 × 0.4
3. 15.7 × 100
4. 2.8 ÷ 10
5. 25 × 1.4
6. 18 ÷ 7 (Answer to 3 decimal places.)
7. 6.71 − 0.8

8. $9.62 \div 0.4$
9. $0.7 \div 7$
10. $191.2 \div 1.8$ (Answer to 3 decimal places.)
11. $5.2 + 0.3 + 12.75$ (RSA)
12. 13.6×1.5 (RSA)
13. 0.86×1000 (RSA)
14. $17.65 \div 100$ (RSA)
15. $76.4 + 7.6$ (RSA)
16. 2.65×3 (RSA)
17. $9.72 \div 6$ (RSA)
18. $0.6 + 60.66 + 6.6$ (RSA)
19. 3.14×8.5 (RSA)
20. $28.39 \div 100$ (RSA)

UNIT THREE

A

Write the name of the shaded fraction:

1. 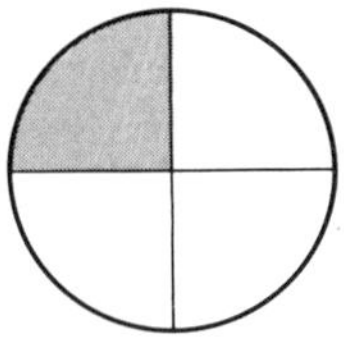

2. 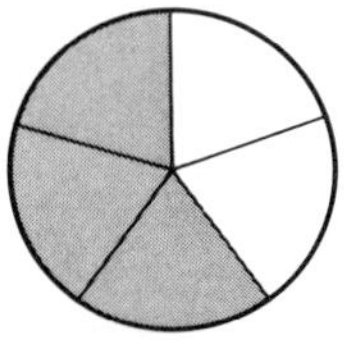

3. 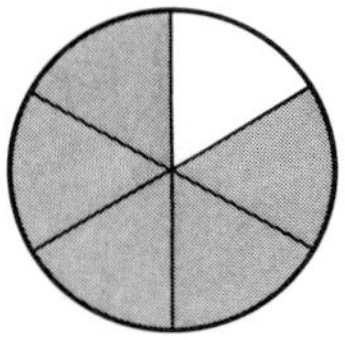

4. 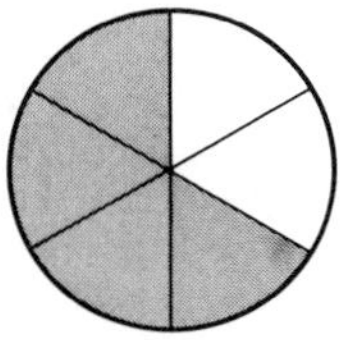

Draw a diagram to show these fractions:

5. $\frac{3}{4}$
6. $\frac{2}{5}$
7. $\frac{5}{6}$
8. $\frac{2}{3}$

9. Share these three cakes equally between four people. How much do they get each?

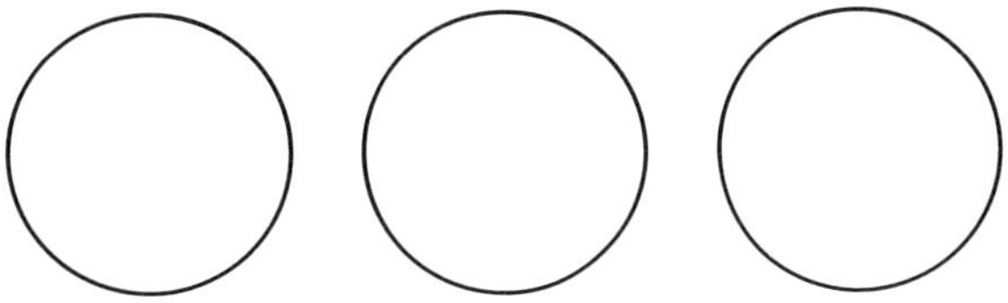

10. Share these four cakes equally between six people. How much do they get each?

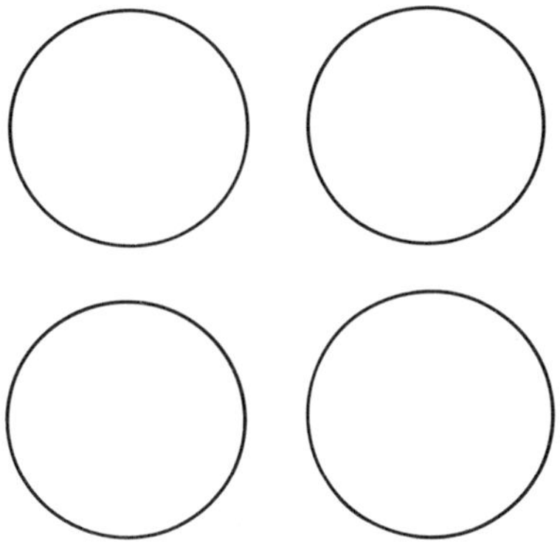

B

Simplify:

1. $\frac{5}{10}$
2. $\frac{8}{12}$
3. $\frac{9}{15}$
4. $\frac{4}{6}$
5. $\frac{8}{16}$
6. $\frac{5}{15}$
7. $\frac{9}{21}$
8. $\frac{4}{12}$
9. $\frac{18}{24}$
10. $\frac{36}{68}$

Complete the following:

11. $\frac{2}{3} = \frac{}{6}$
12. $\frac{3}{4} = \frac{}{12}$
13. $\frac{1}{3} = \frac{}{9}$
14. $\frac{11}{12} = \frac{}{36}$
15. $\frac{2}{5} = \frac{}{20}$
16. $\frac{1}{4} = \frac{}{20}$
17. $\frac{2}{3} = \frac{}{9}$
18. $\frac{1}{7} = \frac{}{21}$
19. $\frac{3}{5} = \frac{}{30}$
20. $\frac{3}{4} = \frac{}{80}$

Change to Improper Fractions:

21. $5\frac{1}{2}$
22. $3\frac{1}{3}$
23. $4\frac{1}{2}$
24. $1\frac{2}{3}$
25. $6\frac{1}{4}$
26. $4\frac{1}{3}$
27. $9\frac{1}{4}$
28. $10\frac{1}{5}$
29. $12\frac{1}{3}$
30. $21\frac{1}{6}$

Change to Mixed Numbers:

31. $\frac{11}{4}$
32. $\frac{15}{3}$
33. $\frac{18}{7}$
34. $\frac{16}{5}$
35. $\frac{21}{4}$
36. $\frac{25}{3}$
37. $\frac{33}{4}$
38. $\frac{38}{5}$
39. $\frac{27}{8}$
40. $\frac{93}{10}$

C Calculate:

1. $2\frac{1}{2} + 1\frac{1}{2}$
2. $3\frac{1}{4} + 1\frac{1}{2}$
3. $4\frac{1}{5} + 1\frac{1}{3}$
4. $9\frac{1}{2} + 1\frac{3}{4}$
5. $3\frac{2}{3} + 2\frac{2}{3}$
6. $8\frac{1}{12} + 4\frac{3}{4}$
7. $9\frac{2}{3} + 3\frac{2}{9}$
8. $3\frac{2}{5} + 4\frac{2}{3}$
9. $112\frac{1}{9} + 57\frac{2}{3}$
10. $28\frac{4}{7} + 19\frac{2}{9}$

D Calculate:

1. $3\frac{1}{2} - 1\frac{1}{4}$
2. $9\frac{2}{3} - 4\frac{1}{2}$
3. $6\frac{1}{7} - 1\frac{2}{3}$
4. $9\frac{1}{3} - 4\frac{3}{4}$
5. $3\frac{3}{4} - 2\frac{2}{3}$
6. $8\frac{2}{3} - 4\frac{3}{8}$
7. $\frac{15}{16} - \frac{3}{8}$
8. $8\frac{11}{12} - 4\frac{3}{4}$
9. $12\frac{1}{9} - 3\frac{1}{7}$
10. $206\frac{1}{4} - 115\frac{2}{9}$

E Calculate:

1. $1\frac{1}{2} \times 1\frac{1}{2}$
2. $3\frac{1}{3} \times 4\frac{1}{2}$
3. $6\frac{2}{3} \times 1\frac{3}{5}$
4. $3\frac{3}{4} \times 3\frac{1}{5}$
5. $12\frac{1}{2} \times \frac{1}{5}$
6. $3\frac{3}{5} \times 4\frac{5}{6}$
7. $7\frac{1}{3} \times \frac{9}{11}$
8. $1\frac{2}{3} \times 1\frac{7}{20}$
9. $\frac{5}{21} \times \frac{14}{15}$
10. $6\frac{1}{4} \times 1\frac{1}{5}$

F Calculate:

1. $1\frac{1}{2} \div 1\frac{1}{2}$
2. $1\frac{1}{3} \div 2\frac{2}{3}$
3. $\frac{7}{9} \div \frac{14}{15}$
4. $2\frac{2}{5} \div 2\frac{7}{10}$
5. $1\frac{3}{4} \div 2\frac{5}{8}$
6. $3\frac{3}{5} \div 2\frac{2}{5}$
7. $9\frac{1}{2} \div 1\frac{1}{3}$
8. $7\frac{2}{3} \div 8\frac{1}{4}$
9. $17\frac{1}{3} \div 1\frac{4}{9}$
10. $9\frac{3}{5} \div 2\frac{2}{15}$

G Change to fractions and simplify:

1. 0.4
2. 0.75
3. 2.4
4. 6.85
5. 9.25
6. 4.125
7. 6.375
8. 21.45
9. 14.325
10. 12.95

H Change to decimals. Work to 3 decimal places:

1. $\frac{3}{4}$
2. $\frac{3}{8}$
3. $\frac{2}{5}$
4. $\frac{4}{9}$
5. $\frac{2}{3}$
6. $9\frac{3}{5}$
7. $8\frac{3}{7}$
8. $12\frac{4}{5}$
9. $\frac{11}{14}$
10. $2\frac{15}{23}$

I Mixed Examples:

1. Shade $\frac{3}{8}$ of this circle.

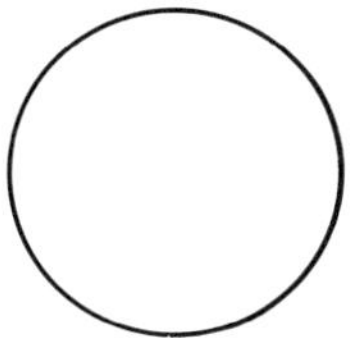

2. $3\frac{1}{3} + 2\frac{3}{4}$
3. Change $\frac{17}{5}$ to a mixed number.
4. Complete $\frac{2}{3} = \frac{}{12}$
5. Change 1.14 to a fraction and simplify.
6. $9\frac{1}{2} - 4\frac{2}{3}$
7. Simplify $\frac{21}{8}$.
8. Change $7\frac{3}{8}$ to a decimal.
9. $4\frac{2}{3} \times 8\frac{1}{5}$
10. $2\frac{3}{4} \div 1\frac{4}{7}$
11. $4\frac{1}{4} - 2\frac{5}{8}$ (RSA)
12. $6\frac{2}{3} \div \frac{5}{6}$ (RSA)
13. $2\frac{1}{4} + 1\frac{5}{8}$ (RSA)
14. $2\frac{1}{2} \div \frac{3}{4}$ (RSA)
15. $3\frac{1}{3} \times 2\frac{1}{2}$ (RSA)
16. $2\frac{2}{5} \times 1\frac{1}{2}$ (RSA)
17. $\frac{3}{4} + \frac{2}{3}$ (RSA)
18. $1\frac{1}{3} \times 1\frac{1}{4}$ (RSA)
19. $2\frac{2}{3} + 3\frac{3}{4}$ (RSA)
20. $3\frac{1}{4} \div 6\frac{1}{2}$ (RSA)

UNIT FOUR

A

Change to pence:

1. £5.12
2. £6.27
3. £12
4. £9.04
5. £8.30

Change to pounds:

6. 125 p
7. 638 p
8. 1312 p
9. 29 p
10. 5 p

B

1. £3.12 – £1.89
2. £8.21 + £4.92 + £14.80
3. £5 – 74 p
4. 2.85 M + 1.76 M
5. 12.82 francs + 8 francs
6. \$12.82 – \$4.95
7. A girl goes shopping. She buys a bag of sugar costing 44 p and a tin of peas costing 38 p. How much change does she get from £1?
8. A meal in a restaurant in France costs 95 francs. A bottle of wine costs 30 francs and a cup of coffee 5.50 francs. Find the total bill.

The next two questions refer to this menu.

Hamburger	55 p
Cheeseburger	65 p
Largeburger	105 p
Chips	36 p
Coffee	34 p
Cola	38 p

9. Find the cost of a hamburger, chips and a cola.
10. A family purchase two hamburgers, one largeburger, three chips, one coffee, and two colas. Find the total cost.

C

1. £3.15 × 6
2. £2.35 × 9
3. £2.44 × $3\frac{1}{2}$
4. 59 p × 14
5. \$3.99 × 8
6. 2.20 francs × 6

7. Find the total cost of six books at £2.25.
8. Find the cost of $4\frac{1}{2}$ metres of material at £6.90 per metre.
9. Find the cost of five boxes of biscuits at £2.15 per box.
10. Find the cost of 15 theatre tickets at $2.95 each.

D

1. £1.89 ÷ 7
2. £2.12 ÷ 2
3. £5.10 ÷ 3
4. £8.00 ÷ 5
5. $9.75 ÷ 15
6. 16.31 francs ÷ 7
7. 22 M ÷ 8
8. £2932 ÷ 8
9. Share £5 equally between four people.
10. Three people decide to buy a present costing £17.61 between them. How much do they each pay?

Use this table for section E and section F.

Tourist rates (amount for one pound)	
France	10.8 francs
Germany	3.5 marks
Italy	2200 lire
Spain	200 pesetas
United States	$1.10

E

1. Change £10 to French francs.
2. Change £50 to German marks.
3. Change £20 to Italian lire.
4. Change £200 to US dollars.
5. Change £40 to Spanish pesetas.
6. Change 21.6 francs to pounds.
7. Change 700 M to pounds.
8. Change 6600 lira to pounds.
9. Change 950 pesetas to pounds.
10. Change $880 to pounds.

F

1. On a trip to France I changed £120 to francs. How many francs did I get?
2. A meal costs 70 M in Germany. What is the equivalent cost in pounds?
3. Change £40 into Italian lire.
4. Change 2500 pesetas into pounds.
5. Which is the greater: $5.50 or £4.50?
6. Change $500 to pounds. (Answer to the nearest pound.)
7. A ferry company allows you to pay for goods in either pounds or francs. A bottle of wine costs 16.20 francs. What is the price in pounds?
8. A book costs £4.50 in Britain. What is the equivalent cost in Spain?
9. A man goes to France with £220. He changes all the money into francs and spends 1800 francs. He changes the remainder back into pounds. How much is left, giving your answer to the nearest pound?
10. An American comes to London with $500. He spends £40 a day for seven days. He also spends £50 on gifts. How many pounds has he left? (Answer to the nearest pound.)

G

Mixed Examples:

1. £5.17 + £2.92 + £7.26
2. £7.15 − £2.92
3. I buy six books costing £2.95 each. How much change do I get from £20?
4. $12.4 × 8
5. Change £230 to francs (£1 = 10.8 francs).
6. Find the total cost of four pens at £1.20 each and seven pencils at 23 p each.
7. Find the cost of 24 bottles of lemonade at 27 p per bottle.
8. How many bottles of milk costing 21 p each can I buy for £5?
9. A meal on a French ferry is priced at £5.50 or 60 francs. Which is the cheapest price? (£1 = 10.8 francs.)

10. A girl spends 52 p each day on bus fares to school. She goes to school five days a week, 40 weeks per year. Find the total amount she spends on fares.
11. £8.72 × 9 (RSA)
12. £16.45 ÷ 7 (RSA)
13. Convert £9 to pesetas, if one pound equals 187.50 pesetas. (RSA)
14. If 17 articles cost £62.56 find the cost of one article. (RSA)
15. A person earns £94 per week. What is his annual salary (52 weeks) to the nearest £100. (RSA)
16. If £1 = $1.59, how many dollars will I get when I change £250? (RSA)
17. A young couple planning to marry save £67.50 each month. How much will they save in 3 years? (RSA)
18. Fish and chips cost 54 p. The chips cost 18 p. What will be the cost of fish and chips three times and two fish? (RSA)
19. How many packets of screws at 29 p each can be bought for £1.60? (RSA)
20. The rate of exchange of pounds sterling to West German Deutschmarks is £1.00 = 4.04 M. How many Deutschmarks can be obtained for £115.00? (RSA)

UNIT FIVE

A
Fill in the gaps:

1. 1 m = cm
2. 1 kg = g
3. 1 ℓ = cℓ
4. 1000 m = 1
5. 1000 mℓ = 1
6. 1000 g = 1
7. 1 km = m
8. 1 m = 1000
9. 1 ℓ = 100
10. 1 m = 100

B
Fill in the gaps:

1. 1 yd = ft
2. 12 in = 1
3. 16 = 1 lb
4. 1 gall = pt
5. 1 yd = in
6. 1760 yd = 1
7. 1 lb = oz
8. 8 pt = 1
9. 1 ft = in
10. 1 mile = yd

C
Fill in the gaps:

1. 2.8 cm = mm = cm mm
2. 12.5 kg = g = kg g
3. 2412 m = km = km m
4. 8620 mℓ = ℓ = ℓ mℓ
5. 2380 cm = m = m cm
6. 25 in = ft in
7. 17 ft = yd ft
8. 73 oz = lb oz
9. $5\frac{1}{2}$ gal = gal pt = pt
10. $1\frac{3}{4}$ lb = lb oz = oz

D

1. 2 m 15 cm + 4 m 92 cm (Answer in metres and centimetres.)
2. 3.1 m + 2.5 m (Answer in metres.)
3. 8.5 m + 50 cm + 75 cm (Answer in metres.)
4. 2.5 kg + $3\frac{1}{4}$ kg + 750 g (Answer in kilograms.)
5. 250 g + 550 g + 890 g + 2140 g (Answer in kilograms.)
6. $8\frac{1}{2}$ ℓ – 750 cℓ (Answer in litres.)
7. 12 kg – 8.5 kg (Answer in grams.)
8. A wooden box weighs 1.5 kg when empty. Find the weight of the box filled with four cauliflowers weighing 700 g, 850 g, 490 g, and 770 g. (Answer in kilograms.)
9. A girl has 6 m of material. She gives 1 m 40 cm to a friend and uses 2 m 80 cm herself. How many centimetres are left?
10. A bottle contains 2 ℓ of cola. If I drink 250 mℓ and my friend drinks 350 mℓ, how much is left? (Answer in millilitres.)

E

1. 2.4 ℓ × 8
2. 5.8 kg × 4
3. 8 ℓ ÷ 16 (Answer in centilitres.)
4. 2 km ÷ 5 (Answer in metres.)
5. 250 g × 14 (Answer in kilograms.)
6. 24 mm × 9 (Answer in centimetres.)
7. A farmer has 12 pieces of fence each 1.5 m long. Find the total length.

8. A relay team of eight runners have to run 15 km between them. How far should they each run? (Answer in metres.)
9. A grocer buys 144 packs of butter each weighing 250 g and 36 packs each weighing 500 g. How many kilograms of butter has he altogether?
10. I have 5 ℓ of lemonade. Forty children are thirsty. How much will each receive? (Answer in millimetres.)

F

1. 2 ft 3 in + 4 ft 9 in + 8 in
2. 5 gal ÷ 8 (Answer in pints.)
3. 1 lb 5 oz × 6
4. 1 mile ÷ 5 (Answer in yards.)
5. 7 pt + 5 pt + 3 pt (Answer in gallons and pints.)
6. 8 lb 2 oz ÷ 5 (Answer in pounds and ounces.)
7. 2 lb 6 oz − 1 lb $7\frac{1}{2}$ oz
8. A pencil is 9 in long. How many pencils are needed to reach 5 yd when placed end to end?
9. A man weighs 15 stone 9 lb. He goes on a diet and loses 21 lb. Find his new weight (14 lb = 1 stone.)
10. A family buys 18 pt of milk per week. How many gallons do they buy in a year?

G

1. Change 8 in to centimetres (1 in = 2.54 cm).
2. Change 5 miles to kilometres (1 mile = 1.6 km).
3. Change 9 gal to litres (1 gal = 4.5 ℓ).
4. Change 95 cm to feet and inches (1 cm = 0.4 in).
5. Change 44 lb to kilograms (1 kg = 2.2 lb).
6. Change 90 ℓ to gallons (1 gal = 4.5 ℓ).
7. Change 32 miles to kilometres (1 mile = 1.6 km).
8. Change 2 ft 4 in to centimetres (1 in = 2.54 cm).
9. Change 50 kg to pounds (1 kg = 2.2 lb).
10. Change 100 km to miles (1 mile = 1.6 km).

H

Mixed Examples:

1. Express 3 yd 2 ft in feet.
2. Change 4268 mm to metres.
3. 3.52 m + 36 cm + 2140 mm (Answer in centimetres.)
4. 2 ft 6 in − 1 ft 11 in
5. Change 17 gal to litres (1 gal = 4.5 ℓ).
6. Change 508 cm to feet and inches (1 in = 2.54 cm).
7. 40 kg ÷ 16 (Answer in grams.)
8. A piece of string is 9 m long. I use 153 cm for one parcel and 97 cm for another. How much string is left? (Answer in metres.)
9. 315 mℓ × 8 (Answer in litres.)
10. 34 ℓ ÷ 5 (Answer in centilitres.)
11. 4 m 26 cm × 10 (ans in m and cm.) (RSA)
12. What is the total weight in kg of 300 packets each weighing 165 g? (RSA)
13. How many litres are in 16 bottles each with a capacity of 75 cℓ? (RSA)
14. What is the total length of 27 pieces of wood each 42 cm long? (Give your answer in m and cm.) (RSA)
15. 42 kg ÷ 4 (ans in grams.) (RSA)
16. How many litres of wine will be required to fill 40 bottles if each bottle has a capacity of 70 cℓ? (RSA)
17. In an average family saloon car the shortest stopping distance at 50 mph is 175 feet. Convert the stopping distance to metres and centimetres. (1 foot = 0.3 m) (RSA)
18. A cricket pitch measures 22 yards between the wickets. What is this length in metres? (1 yd = 0.91 m) (RSA)
19. What is the total weight in kg of 150 packets each weighing 225 g? (RSA)
20. A family used 2 pints of milk each weekday and 3 pints on Saturday and 3 pints on Sunday. How many gallons of milk does this family use in 1 year. (1 year = 52 weeks) (RSA)

UNIT SIX

A

Round these prices to the nearest penny:

1. £5.284
2. 86.7 p
3. £10.7496
4. £12.099
5. 92.5 p
6. £0.0412
7. 9.6789 p
8. £7.1746
9. £24.723
10. 62.39 p

B

Round to the number of decimal places shown:

1. 15.912 (2 d.p.)
2. 20.714 (1 d.p.)
3. 0.004 55 (3 d.p.)
4. 19.99 (1 d.p.)
5. 0.0108 (3 d.p.)
6. 27.944 (2 d.p.)
7. 10.002 (1 d.p.)
8. 0.000 42 (1 d.p.)
9. 11.62 (1 d.p.)
10. 72.848 (2 d.p.)

C

Round to the number of significant figures shown:

1. 19.24 (3 sig. fig.)
2. 23.72 (2 sig. fig.)
3. 104.71 (3 sig. fig.)
4. 92.894 (3 sig. fig.)
5. 21.214 (4 sig. fig.)
6. 0.014 67 (3 sig. fig.)
7. 0.028 (2 sig. fig.)
8. 200.614 (3 sig. fig.)
9. 19 784 (3 sig. fig.)
10. 2189 733 (2 sig. fig.)

D

1. Change $1\frac{7}{8}$ to a decimal.
2. Change $1\frac{4}{5}$ to a decimal.
3. Arrange in order of size, smallest first: $1\frac{4}{5}$, $1\frac{7}{8}$, 1.85
4. Complete $\frac{3}{4} = \frac{}{20}$
5. Complete $\frac{4}{5} = \frac{}{20}$
6. Arrange in order of size, largest first: $\frac{3}{4}$, $\frac{17}{20}$, $\frac{4}{5}$
7. Which is the largest of: $9\frac{1}{2}$, 9.51, $\frac{28}{3}$?
8. Arrange in order of size, smallest first: $\sqrt{100}$, 3^2, $\frac{19}{2}$
9. Which is the smallest of: $\frac{9}{4}$, 2.3, $2\frac{1}{5}$?
10. Arrange in order of size, largest first: $3\frac{1}{7}$, 3.14, $3\frac{2}{15}$

E

Find the value of:

1. 9^2
2. $\left(1\frac{1}{2}\right)^2$
3. 2.1^2
4. 30^2
5. 200^2
6. $\sqrt{25}$
7. $\sqrt{64}$
8. $\sqrt{10\,000}$
9. $\sqrt{0.25}$
10. $\sqrt{\frac{1}{9}}$

F

Find the HCF of each group of numbers:

1. 12, 16
2. 36, 48
3. 5, 6
4. 72, 84
5. 90, 100
6. 11, 22, 44
7. 72, 48, 80
8. 100, 81, 36
9. 15, 20, 30
10. 14, 28, 35

G

Find the LCM of each group of numbers:

1. 2, 4
2. 4, 6
3. 7, 8
4. 4, 8
5. 15, 20
6. 3, 5, 6
7. 15, 12, 8
8. 11, 2, 4
9. 5, 10, 3
10. 4, 7, 28

H

Mixed Examples:

1. Round £9.487 to the nearest penny
2. Round 8.7204 to 3 significant figures.
3. Round 8.7204 to 3 decimal places.
4. Find the smallest of $\sqrt{25}$, $\frac{21}{4}$, 4.99
5. Find the LCM of 14, 7, 2
6. Find the HCF of 12, 15, 21
7. Divide £12.65 by 3 and give the answer to the nearest penny.
8. 7.21 × 4.3 (round the answer to 2 decimal places).
9. Find $\sqrt{144}$
10. Find $(2.6)^2$

11. Arrange the following numbers in order of size, largest first: $3\frac{5}{9}$, $\frac{11}{3}$, 3.6 (RSA)
12. Find the LCM of 14 and 21 (RSA)
13. Write the following numbers in order of size, largest first: $2\frac{2}{9}$, $\frac{8}{3}$, 2.6 (RSA)
14. A company representative travels 290 miles per month in his car. What is his annual mileage to the nearest 100 miles? (RSA)
15. Put the following numbers in order of size, largest first: $\frac{13}{5}$, 2.7, $2\frac{3}{4}$ (RSA)
16. Find the square root of $7\frac{1}{9}$ (RSA)
17. Find the LCM of 4, 10, and 12 (RSA)
18. Find the value of $\left(3\frac{1}{2}\right)^2$ (RSA)
19. Express 19 479 to the nearest thousand (RSA)
20. Find the square root of 1.21 (RSA)

UNIT SEVEN

A Change these fractions to percentages:

1. $\frac{5}{8}$
2. $\frac{2}{5}$
3. $\frac{7}{8}$
4. $\frac{2}{3}$
5. $\frac{3}{10}$
6. $2\frac{1}{2}$
7. $\frac{3}{4}$
8. $\frac{9}{10}$
9. $\frac{7}{20}$
10. $\frac{23}{50}$

B Change these percentages to fractions:

1. 80%
2. 50%
3. 45%
4. 16%
5. 125%
6. $12\frac{1}{2}$%
7. $27\frac{1}{2}$%
8. $8\frac{1}{3}$%
9. $133\frac{1}{3}$%
10. 350%

C

1. A girl scores 17 out of 20 in a test. Express this as a percentage.
2. Another girl scores 34 out of 50. Express this as a percentage.
3. Express £1.25 as a percentage of £5.
4. Express 250 g as a percentage of 4 kg.
5. Express 50 cℓ as a percentage of 2.5 ℓ.
6. A bottle contains 2 ℓ of wine. The waiter spills 20 mℓ. What percentage is spilt?
7. A bag of crisps goes up from 10 p to 12 p. Express the increase as a percentage of the original price.
8. The price of a 15 p chocolate bar goes up by 2 p. Express this increase as a percentage.
9. A bag contains 1 kg of sugar. I use 700 g. What percentage is left?
10. A car costs £4500. A man has £3000. What percentage of the cost does he still need to save?

D

1. Find 5% of £200.
2. Find 8% of £400.
3. Find $12\frac{1}{2}$% of 8 kg.
4. Find 25% of 6 ℓ. (Answer in litres and millilitres.)
5. Find 75% of 20 m.
6. Increase £4 by 20%.
7. Decrease £12 by 25%.
8. A meal costs £15, plus 15% VAT. Find the total price.
9. A test is marked out of 60. How many must I get to pass if the pass mark is 45%?
10. A tank will hold 1000 ℓ of oil. If 23% has been used, how much is left?

E

1. Find the Simple Interest on £200 invested for 2 years at 10%.
2. Find the Simple Interest on £1200 invested for 3 years at 8%.
3. Find the Simple Interest on £5000 invested for 4 years at 12%.

4. Find the Simple Interest on \$1500 invested for 10 years at 5%.
5. Find the Simple Interest on 5000 marks invested for 1 year at 9%.
6. Find the Simple Interest on £100 invested for 6 months at 8%.
7. Find the Simple Interest on £400 invested for $2\frac{1}{2}$ years at 10%.
8. Find the Simple Interest on £2000 invested for 2 years at $12\frac{1}{2}$%.
9. A man borrows £5000 for 1 year and has to pay 18% interest. How much does he pay back altogether?
10. A girl borrows £5 from her friend and agrees to pay 2% Simple Interest per week. She pays the money back after six weeks. How much interest does she pay?

F

1. A shop offers £100 worth of clothes for 12 monthly payments of £10. Find the total price paid for the goods.
2. A washing machine costs £204 cash or 24 monthly payments of £8.95. Find the extra cost of the monthly scheme.
3. A television is offered for £90 deposit and 36 monthly payments of £9.00. Find the total cost.
4. A bicycle costs £20 deposit and 18 monthly payments of £4.99. Find the total price.
5. A car costs £7200 or one fifth deposit and 36 monthly payments of £190. Find the total Hire Purchase price.
6. A cooker costs £400 cash or one quarter deposit and 12 monthly payments of £30. Find the extra cost of HP and express this as a percentage of the cost price.
7. A radio is advertised as £19 cash or £5 deposit and six monthly payments of £2.50. Find the extra cost of HP and express this as a percentage of the cost price. (Answer to the nearest per cent.)
8. A builder says he will charge £400 to do a job or £200 now and £20 a month for a year. Express the extra cost as a percentage of the cost price.
9. A dress costs £15.99 cash or 20 weekly payments of 80 p. Find the extra cost of buying the dress by weekly payments.
10. A computer costs £1200. A dealer offers it for £300 now and 36 monthly payments of £30. Express the extra cost as a percentage of the cost price.

G Mixed Examples:

1. Change $1\frac{3}{8}$ to a percentage.
2. Change 85% to a fraction.
3. Express 24 p as a percentage of £3.
4. Find 15% of £300.
5. Increase 30 kg by 20%.
6. Find the Simple Interest on £200 invested for 3 years at 8%.
7. Decrease £4.50 by 20%.
8. A boat costs £1200 or £400 deposit and 12 monthly payments of £70. Express the extra cost as a percentage of the cash price.
9. A coat can be purchased for 30 weekly payments of £1.99. Find the total cost.
10. A gas fire costs £180 or 10% deposit and 8 quarterly instalments of £25.25. Express the extra cost as a percentage. (Answer to the nearest per cent.)
11. What is $7\frac{1}{2}$% of £14? (RSA)
12. Express $\frac{7}{8}$ as a percentage. (RSA)
13. Express 16 cm as a percentage of 4 m. (RSA)
14. A hire purchase company charges 15% per annum Simple Interest. How much interest will be charged over 3 years for goods valued at £450? (RSA)
15. Find the total cost of a hand brush at 98 p after adding 15% VAT to the price of the brush. (Answer to the nearest penny.) (RSA)
16. £586.00 was invested in a savings bank at 5% Simple Interest per annum. Calculate the amount in the account after 1 year. (RSA)

17. A company with sales totalling £32 000 has assets of £800 000. Find the percentage of the company's total sales to the company's assets. (RSA)
18. A car costing £4500 is sold 2 years later for £3300. What is the percentage loss on this sale as a percentage of the original price? (RSA)
19. What is 15% of £4.50? (RSA)
20. Express £1.50 as a percentage of £7.50. (RSA)

UNIT EIGHT

A

1. Find the average of 12, 17, 21, 29, 36.
2. Find the average of 150, 152, 154, 159.
3. Find the average cost of six items costing £1.25, £1.30, £1.40, £1.62, £1.93, £2.04.
4. Find the average weight of five parcels weighing 1.2 kg, 700 g, 1.34 kg, 250 g, 570 g.
5. The temperatures recorded during a week's holiday are 19°, 17°, 20°, 21°, 18°, 17°, 20°. Find the average temperature. (Answer to 1 decimal place.)
6. Four athletes each run 1000 m. Their times are 2 min 58 s, 2 min 48 s, 2 min 50 s and 3 min 0 s. Find their average time.
7. Find the average cost of six 'Chocbars' at 15 p each and four 'Toffbars' at 13 p each.
8. Find the average length of three pieces of wood of length 24.3 cm, 19.7 cm, and 28.9 cm.
9. Find the average of 0.7, 0.81, 0.85, 0.88, 0.92, 0.98. (Answer to 3 decimal places.)
10. Five new houses in a road are priced at £43 500, £44 500, £45 000, £46 000 and £48 000. Find the average price.

B

Mixed Revision Examples:

1. $27 + 14 + 9 + 135$ (RSA)
2. 52×13 (RSA)
3. $4.8 + 0.06 + 14.59$ (RSA)
4. 6.05×1.5 (RSA)
5. $3\frac{1}{4} \div 6\frac{1}{2}$ (RSA)
6. $2\frac{1}{3} + 4\frac{3}{8}$ (RSA)
7. 0.74×1000 (RSA)
8. If 8 pies cost £19.60, find the cost of 5 pies. (RSA)
9. Express 0.65 as a fraction in its lowest terms. (RSA)
10. Work out the Simple Interest on £750 for $2\frac{1}{2}$ years at 8% (RSA)
11. Find the average of the following amounts of money: £2.35, £5.50, £4.37, £6.45 and £2.33 (RSA)
12. A camera is sold for £96 cash or $\frac{1}{3}$ down payment and 24 monthly payments of £4.38. What is the cost of the camera using these hire purchase terms? (RSA)
13. Find the average of 5, 8, 12, 19, 21 and 25. (RSA)
14. Express 1.5 m as a percentage of 6 m. (RSA)
15. Calculate the Simple Interest on £350 invested for 3 years at 12% p.a. (RSA)
16. A person saved £6.25 regularly each month. How much will be saved in 24 months? (RSA)
17. If 12 apprentices each earn £45 a week and 18 workmen each earn £60 a week, calculate their average weekly wage. (RSA)
18. A package containing boxes of matches is priced at £7.20. Each box is priced at 6 p and each box contains 55 matches. How many matches are there in the package? (RSA)
19. A five-a-side football team is entered in a competition. Find the average age of the team in decimals if the ages of the players are as follows: 17, 18, 19, 21 and 21 years. (RSA)
20. In the Shell Mileage Marathon the 'Thrifter' travelled 2575 miles on 1 gallon of petrol. How far would the vehicle travel on 1 pint of petrol if the rate were maintained? (Answer to the nearest mile) (RSA)

UNIT NINE

A

1. If four books cost £6, what will be the cost of nine similar books?
2. If five cakes cost £7.50, what will three cakes cost?
3. 9 lb potatoes cost 99 p. Find the cost of 12 lb.
4. 12 rolls cost 76 p. What will 18 rolls cost?
5. If six people eat nine loaves of bread at a picnic, how many loaves would you expect ten people to eat?
6. If three sacks of coal last for four weeks, how long will four sacks last?
7. If a boy is paid £4.50 for three hours work, what should he be paid for five hours?
8. If a rail journey costs £25.50 for five people, how much will it cost for seven people?
9. If I need three cans of soup for five people, how many similar cans will I need for 15 people?
10. A theatre takes £155 on Monday when 62 people buy tickets. How much will they take on Saturday when 214 people buy tickets at the same rate?

B

Express these ratios in the lowest terms:

1. 15 : 20
2. 9 : 21
3. £12 : £18
4. 25 kg : 20 kg
5. £450 : £950
6. 2 ℓ : 300 cℓ
7. 750 g : 2 kg
8. 2 lb 4 oz : 3 lb 8 oz
9. 2 ft 8 in : 3 ft 8 in
10. 9 gal : 36 pt

C

1. Divide £20 in the ratio 2 : 3.
2. Divide £100 in the ratio 3 : 1.
3. Divide £36 in the ratio 4 : 5.
4. Divide £72 in the ratio 1 : 2.
5. Divide £108 in the ratio 7 : 5.
6. Share 50 sweets between three children in the ratio 1 : 2 : 2.
7. Share 150 stamps between two boys in the ratio 7 : 8.
8. A mixture consists of eggs, sugar and butter in the ratio 2 : 3 : 1. How much butter is there in $1\frac{1}{2}$ kg of the mixture?
9. A man leaves £35 000 to his four sons in the ratio 2 : 3 : 4 : 5. How much is the largest share?
10. A grocer sells tea from Sri Lanka, India and China in the ratio 5 : 2 : 1. If he is going to order 1400 kg of tea altogether, how much Indian tea should he order?

D

1. Two boys share some marbles in the ratio 2 : 3. If the first boy gets 40 marbles, how many will the second boy get?
2. Pastry is made from flour and fat in the ratio 2 : 1. How much flour do I need to make 300 g of pastry?
3. Two people buy a business. A pays £7000 and B pays £14 000. In the first year they make £3000 profit. How much should they each receive?
4. Two men own a vegetable plot. They agree to share the produce in the same ratio as the number of hours they work. The first man works 36 hours and the second 60 hours. How should they divide 16 lb of runner beans?
5. Three children share some apples in the ratio 2 : 3 : 5. If the first child gets 16 apples, how many do the others get?
6. A girl sorts her records into singles, twelve-inch singles and albums. She finds they are in the ratio 5 : 2 : 1. If she has 24 twelve-inch singles, how many has she altogether?
7. A class consists of 24 boys and 20 girls. What is the ratio of boys to girls?
8. Share 27 gal of petrol in the ratio 1 : 2 : 3.

9. Three children do a paper round between them. Alan delivers 50 papers, Barbara delivers 55 and Charles delivers 65. They receive a total of £2.04. How should they share the money?

10. A breakfast cereal is a mixture of oats, wheat-flakes, fruit and nuts in the ratio 4 : 3 : 2 : 1. What weight of fruit is there in a 750 g packet?

E

1. Find the cost of a mixture of seeds in the ratio 3 : 2 if the first costs 25 p per pound and the second 15 p per pound?
2. A mixture is made from 3 kg tea at £2.20 per kilogram and 2 kg of tea at £1.90 per kilogram. Find the cost of 1 kg of the mixture.
3. A shopkeeper has 20 pencils costing 15 p each, 15 pencils costing 20 p each and 5 pencils costing 32 p each. He mixes them and sells them all at the same price. What price must he charge for each pencil to receive the same amount as before?
4. Two solutions are mixed in the ratio 3 : 4. The first solution costs £1.80 per litre and the second £5.20 per litre. Find the cost of one litre of the mixture. (Answer to the nearest penny.)
5. A mixture is made of sand, cement and gravel in the ratio 1 : 2 : 4. If the cost per 50 kg bag is £2.60, £5.50 and £1.50 respectively, find the cost of 100 kg of the mixture.
6. A mixture of toffees and chocolates is made in the ratio 5 : 4. Toffees cost 44 p for 250 g and chocolates cost 82 p for 250 g. Find the cost of 250 g of the mixture. (Answer to the nearest penny.)
7. A chemist dilutes cough medicine with water in the ratio 1 : 9. If the cough medicine costs £12 per litre and water is free what must he charge for 200 mℓ of the mixture?
8. An engine runs on a mixture of oil and petrol in the ratio 1 : 8. Oil costs 36 p per pint and petrol costs £1.86 per gallon. Find the cost of a gallon of the mixture. (Answer to the nearest penny.)
9. A florist mixes daffodils and tulips in the ratio 3 : 1. Daffodils cost 20 p for ten and tulips cost 50 p for ten. How much will 12 mixed flowers cost?
10. A woman buys brown and white flour. Brown flour costs 75 p for a $1\frac{1}{2}$ kg bag and white flour costs 60 p for a similar bag. She mixes the flour in the ratio 7 : 3. Find the cost of 1 kg of the mixture.

F Mixed Examples:

1. If five books cost £9.00, what will six books cost?
2. Simplify the ratio 500 mℓ: 85 cℓ.
3. Divide £90 in the ratio 2 : 1.
4. Two men divide some money in the ratio 2 : 3. If the first man gets £120, how much will the second man get?
5. Meat and onions are mixed in the ratio 4 : 1 to fill pies. If the meat costs £1.20 per lb and the onions 30 p per lb, find the cost of 1 lb of the mixture.
6. A bus conductor finds the small change in his bag consists of 1 p, 2 p, 5 p and 10 p coins in the ratio 8 : 10 : 7 : 5. If he has 180 coins, how much money is in his bag?
7. Share £1500 between three people in the ratio $2 : 2\frac{1}{2} : 1\frac{1}{2}$.
8. Simplify the ratio 2.5 m : 125 cm.
9. Two men share 500 kg of potatoes. The first man gets 200 kg. In what ratio did they share the potatoes?
10. Carol and Jane share some photographs in the ratio 3 : 5. Carol gets 120 photographs. How many does Jane get?
11. Divide £225 in the ratio 3 : 2. (RSA)
12. Divide 243 in the ratio 5 : 4 (RSA)
13. A music centre is priced at £115.00. The cost of a tape deck, record deck and the speakers can be divided into the ratios 5 : 3 : 2 respectively. Find the cost of the speakers. (RSA)

14. Divide 312 in the ratio 3 : 5. (RSA)
15. Divide £639 in the ratio 5 : 4. (RSA)
16. A sum of money is divided between A and B in the ratio 2 : 3. If A received £16 find the sum of money. (RSA)
17. Three quarters of a number is 60. What is the number? (RSA)
18. Twelve articles cost £66. What is the cost of 5? (RSA)
19. Two thirds of a number is 240. What is the number? (RSA)
20. Divide £84 in the ratio 3 : 4. (RSA)

UNIT TEN

A Find the area of these shapes. Measure in centimetres:

1.

2.

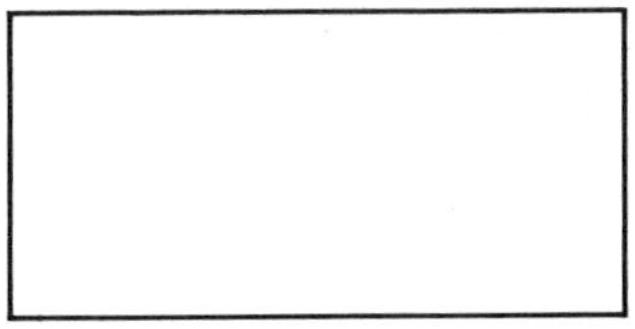

3.

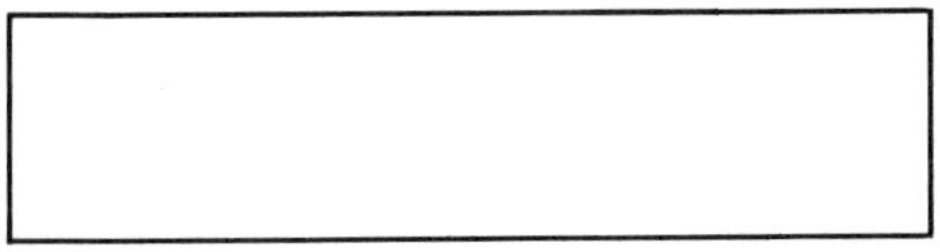

4.

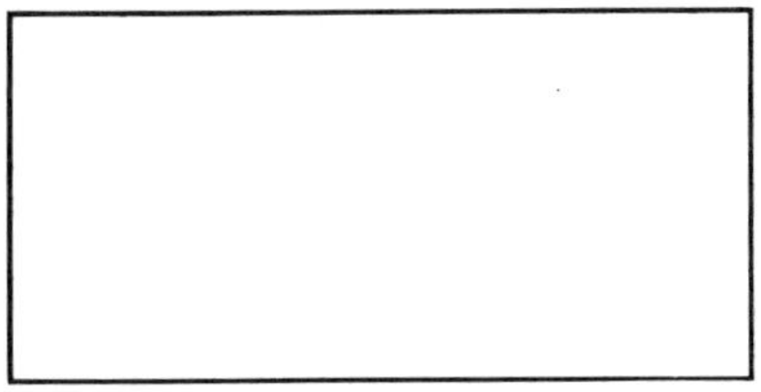

B Fill in the following:

1. 1 cm^2 = mm^2
2. 1 m^2 = cm^2
3. 1 m^2 = mm^2
4. 1 km^2 = m^2
5. 1 ft^2 = in^2
6. 1 yd^2 = ft^2
7. 1600 cm^2 = m^2
8. 18 ft^2 = yd^2
9. 1.028 m^2 = cm^2
10. 275 800 m^2 = km^2

C

1. Find the area of a square of side 4 cm.
2. Find the area of a rectangle 3 in long and 2 in wide.
3. Find the area of a rectangle 9 m long and 4 m wide.
4. Find the area of a rectangle $4\frac{1}{2}$ cm long and $2\frac{1}{2}$ cm wide.
5. Find the area of a carpet 4 m long and 3.4 m wide.
6. Find the area of a painting 3 ft 1 in long and 2 ft 4 in wide. (Answers in square inches.)
7. Find the area of a rectangle 50 cm wide and $2\frac{1}{2}$ m long. (Answer in square metres.)
8. Find the area of a square of side $3\frac{1}{2}$ in.
9. Find the area of a square of side 8.4 mm.
10. Find the area of a rectangle 90 cm wide and 1.2 m long. (Answer in square centimetres.)

D Measure and find the area of these triangles:

1. Measure in centimetres.

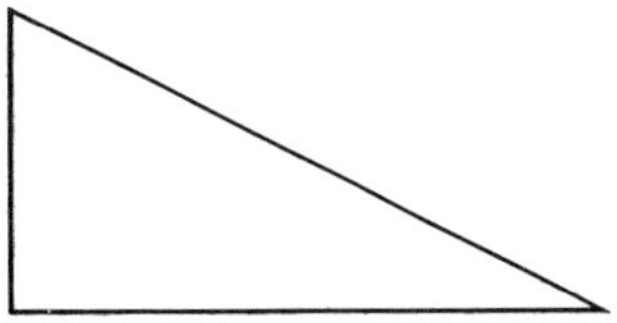

2. Measure in inches.

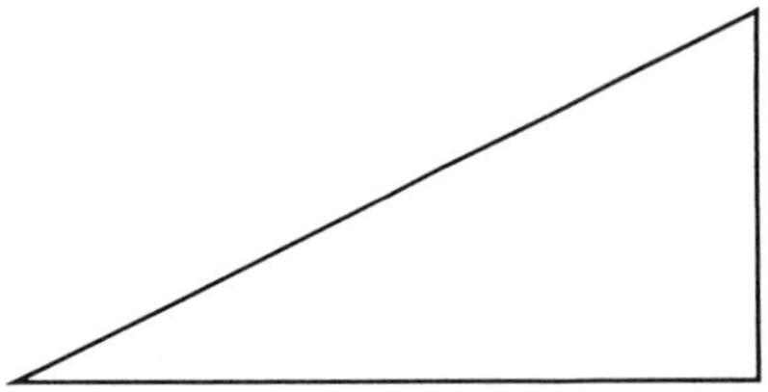

3. Measure in centimetres.

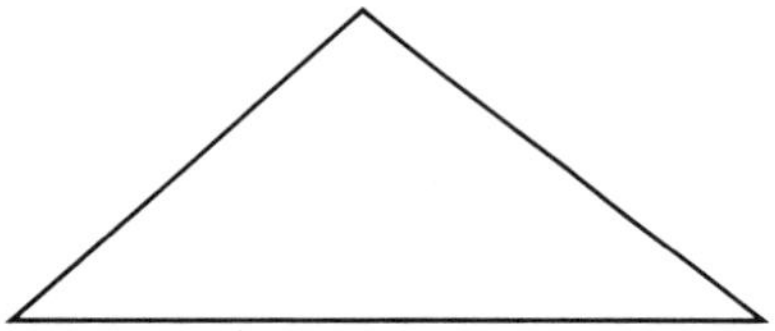

4. Measure in inches.

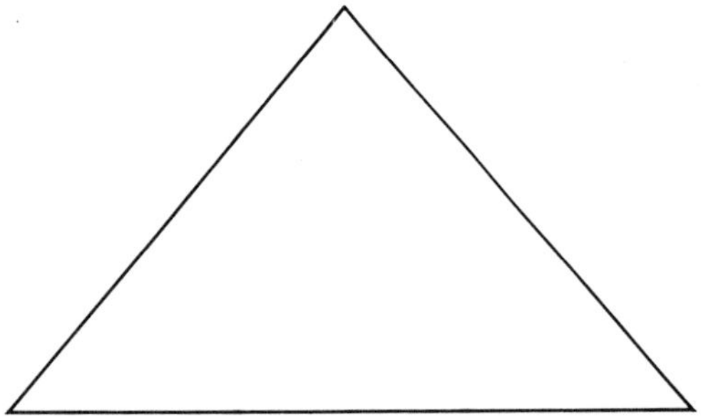

Find the area of these triangles:

5.

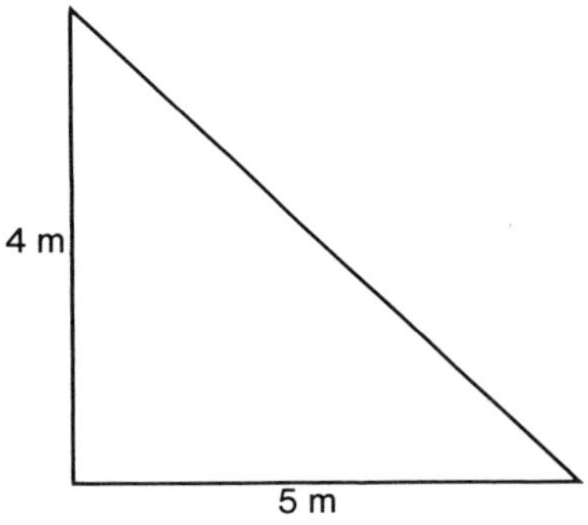

6.

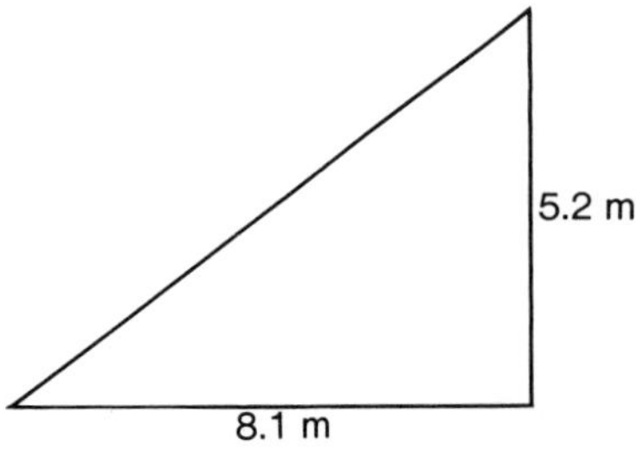

7.

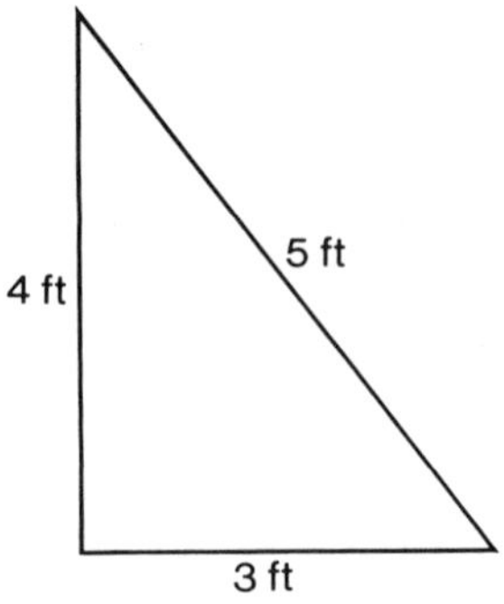

8.

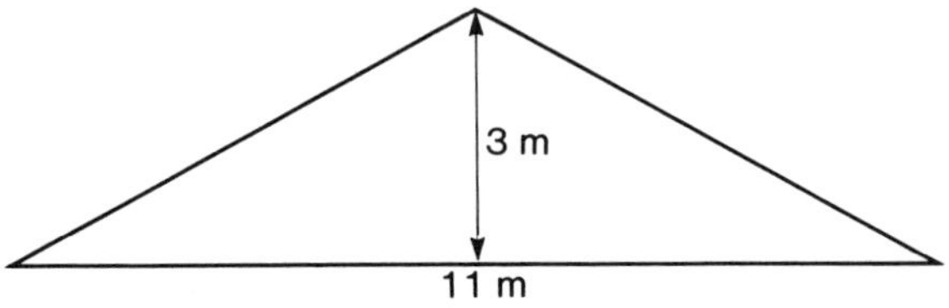

9. A triangular window has a base of 90 cm and a height of 80 cm. Find its area.

10. Find the area of this pirate flag.

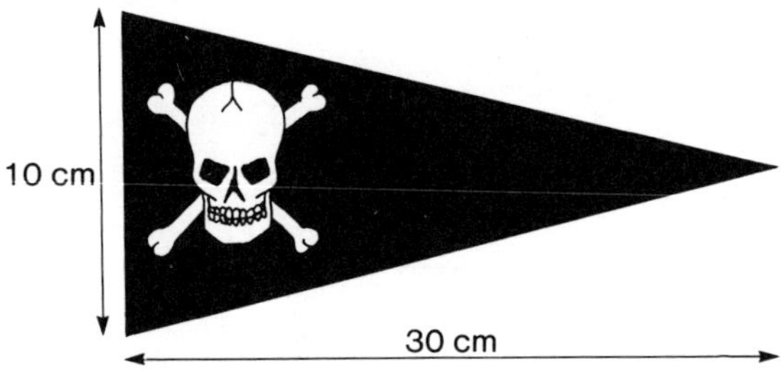

E Find the area of the following circles:

$\pi = \frac{22}{7}$

1. radius 7 m
2. radius 14 cm
3. radius 28 mm
4. diameter 7 cm
5. diameter 140 m
6. radius 2 m
7. radius 5 cm
8. radius 1.5 cm
9. diameter 8 in
10. diameter 6 ft

F

1. If the area of a square is 9 cm^2 find the length of a side.
2. If the area of a rectangle is 20 cm^2 and the width is 4 cm, find the length.

3. A sheet of paper is 10 cm wide and has an area of 275 cm^2. Find its length.
4. A triangle has a base of 4 cm. Its area is 12 cm^2. Find the height.
5. A gardener wants to make a triangular flower bed covering 18 m^2. He wants it to be 6 m from the base to the top. How long must he make the base?
6. A circular rug has an area of 154 m^2. Find the radius $\left(\pi = \frac{22}{7}\right)$.
7. A tin of paint will cover 50 m^2. If a fence is 2 m high and is to be painted on both sides, what length of fence can be painted with one tin?
8. A cake recipe requires a 16 cm square tin. What diameter circular tin has the same area. ($\pi = 3.14$.) (Answer to the nearest centimetre.)
9. A tile is 6 in square. How many will I need to cover a wall 8 ft by 2 ft?
10. How many circles of radius 10 cm can I cut from a piece of wood 40 cm by 80 cm? ($\pi = 3.14$)

G Mixed Examples:

1. Convert 2 m^2 to square centimetres.
2. Convert 5 ft^2 to square inches.
3. Find the area of a rectangle $3\frac{1}{2}$ cm by $8\frac{1}{2}$ cm.
4. Find the area of a circle of radius $2\frac{1}{3}$ cm $\left(\pi = 3\frac{1}{7}\right)$.
5. Find the area of a triangle with base 5 m and height $2\frac{1}{2}$ m.
6. A floor has an area of 150 m^2. The width is 12 m. How long is it?
7. A circle has a diameter of 2.8 m. Find its area ($\pi = 3.14$).
8. Change 520 000 m^2 to square kilometres.
9. A room has an area of 15 m^2. The length is 4 m. Find the width.
10. Find the area of a circle which has a radius of 49 cm $\left(\pi = \frac{22}{7}\right)$.
11. What is the area of a circle which has a radius of 42 cm? $\left(\pi = \frac{22}{7}\right)$ (RSA)
12. A triangular sheet of metal has a base of 3 m and an area of 9.6 m^2. What is its perpendicular height? (RSA)
13. What is the area of a circle of diameter 28 cm? $\left(\pi = \frac{22}{7}\right)$ (RSA)
14. A triangular sheet of metal has an area of 15.6 cm^2 and a base of 5.2 cm. Calculate the perpendicular height. (RSA)
15. Find the cost of fully carpeting a room 4 yards by 3 yards at £5.00 a square yard. (RSA)
16. Find the cost of paving a patio containing 144 square-metre slabs at £5.00 per slab.
17. A circular swimming pool measures 12 metres in diameter. What is the area of the pool? ($\pi = 3.14$) (RSA)
18. Calculate the area of a circle of radius 7 cm. Take $\pi = \frac{22}{7}$. (RSA)
19. A triangle has an area of 60 cm^2. If the base is 10 cm, calculate the perpendicular height. (RSA)
20. Find the area of a circle radius $3\frac{1}{2}$ cm. $\left(\pi = 3\frac{1}{7}\right)$ (RSA)

UNIT ELEVEN

A Find the perimeter of these shapes. Measure in centimetres:

1.

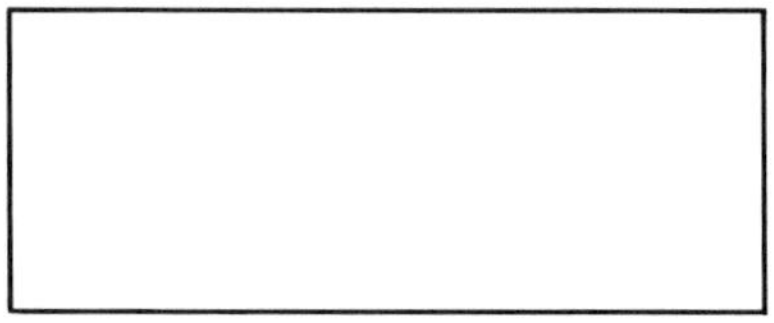

2.

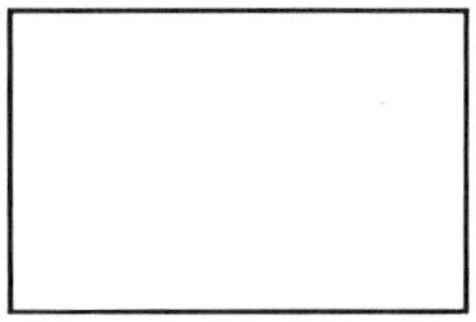

3.

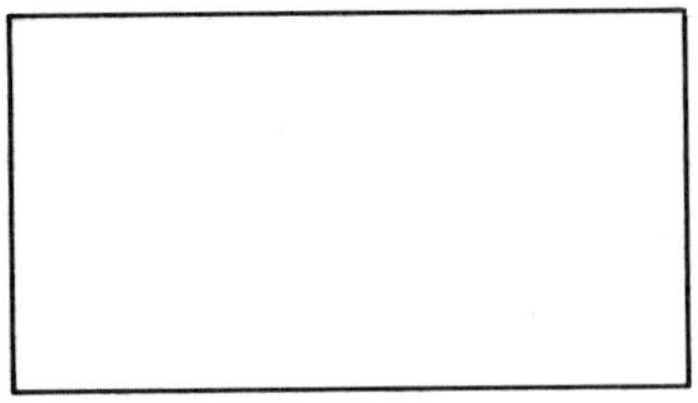

4.

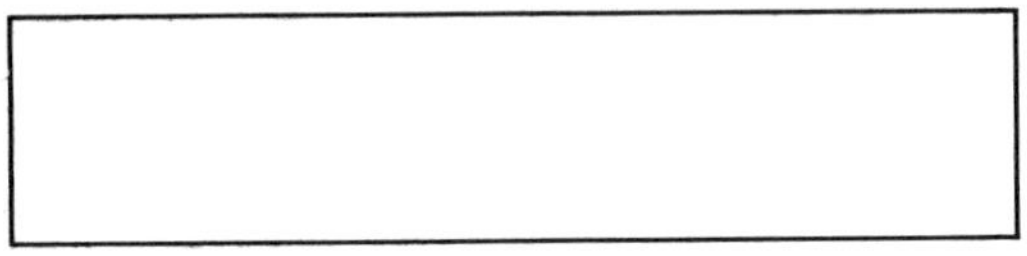

Find the perimeter of these shapes:

5.

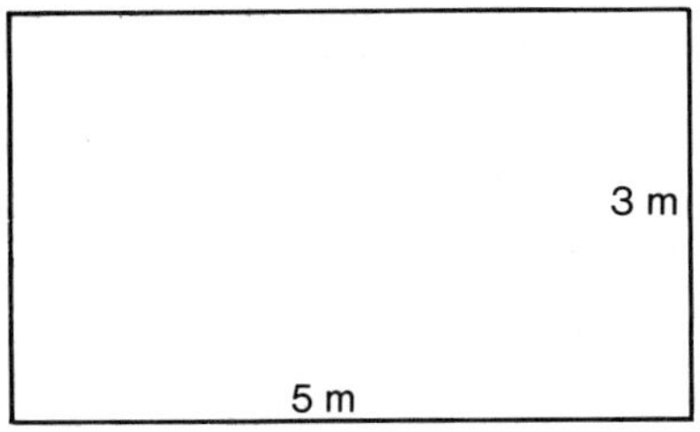

6.

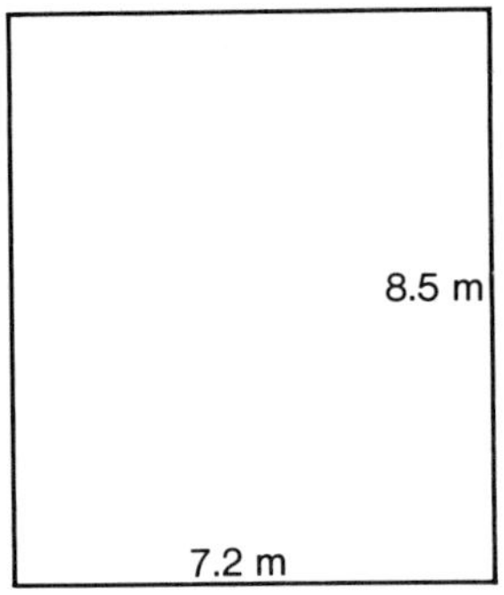

7.

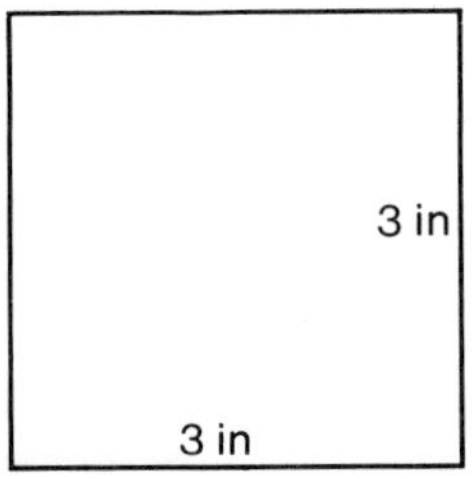

8.

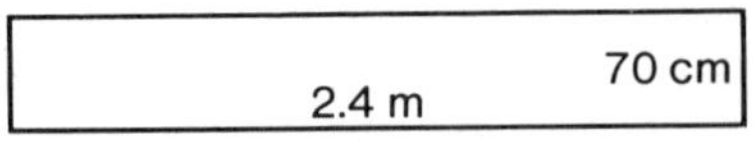

9. A bedroom measures 11 ft 2 in by 8 ft 4 in. Find its perimeter in feet and inches.

10. What is the length of framing tape needed to go round a picture measuring 20 cm by 25 cm?

B

1. The perimeter of a square is 16 m. Find its area.

2. The perimeter of a rectangle is 30 cm. The width is 5 cm. Find its area.

3. The area of a square is 25 in^2. Find its perimeter.

4. The area of a rectangle is 40 m^2. The length is 8 m. Find the perimeter.

5. A carpet measures 4 m by 2.8 m. How much tape is required to stick the edges?

6. A farmer has enough wire fence to make a rectangle 10 m by 20 m. Instead he makes a square. Find the area of the square and the area of the rectangle.

7. A garden measures 30 ft by 20 ft. The owner buys an extra plot of land to increase it to a 30 ft square. How much extra fence does he need to buy? Find the percentage increase in area.

8. The perimeter of a square room is 36 m. Find the cost of carpeting it at £8 per square metre.

9. A square has a side of 5 cm. If I double the perimeter by what percentage will the area increase?

10. A girl has a metre square of silk. She cuts it into four similar squares and binds all the edges with ribbon. What length of ribbon does she use?

C

Find the circumference of the following circles.

$\pi = \frac{22}{7}$ | $\pi = 3.14$

1. radius 7 cm
2. radius 14 cm
3. radius 28 cm
4. diameter 140 m
5. diameter 7 in
6. radius 2 m
7. radius 6 cm
8. radius 2.5 m
9. diameter 10 in
10. diameter 6 ft

D

1. Find the radius of a circle with circumference 88 cm $\left(\pi = \frac{22}{7}\right)$.
2. Find the radius of a circle with circumference 22 m $\left(\pi = 3\frac{1}{7}\right)$.
3. Find the area of a circle with circumference 176 m $\left(\pi = \frac{22}{7}\right)$.
4. Find the area of a circle with circumference 6.28 m ($\pi = 3.14$).
5. Find the circumference of a circle with area 154 cm^2 $\left(\pi = 3\frac{1}{7}\right)$.
6. Find the circumference of a circle with area 616 m^2 $\left(\pi = \frac{22}{7}\right)$.
7. Find the circumference of a circle with area 12.56 cm^2 ($\pi = 3.14$).
8. Find the area of a circle with circumference 62.8 m ($\pi = 3.14$).
9. A circular rug has a circumference of 22 ft. Find its area $\left(\pi = 3\frac{1}{7}\right)$.
10. A racing track has a circumference of 1000 m. Find its radius. (Answer to the nearest metre.) ($\pi = 3.14$)

E

1. Find the volume of a box 2 cm by 3 cm by 4 cm.
2. Find the volume of a box $2\frac{1}{2}$ cm by 4 cm by 6 cm.
3. Find the volume of a box 3.5 cm by 2.1 cm by 8 cm.
4. Find the volume of a cube of side 3 in.
5. Find the volume of a room 4 m by 3 m by 2.5 m.
6. A cube has a volume of 8 cm^3. Find the length of a side.
7. A rectangular box has a volume of 20 cm^3. It is 5 cm long and 2 cm wide. Find its height.
8. A room has a volume of 30 yd^3. The floor measures 9 ft by 12 ft. Find the height. (Answer in feet and inches.)
9. A box is to contain 10 ℓ of wine. The base measures 25 cm square. Find the height of the box. (1 ℓ = 1000 cm^3)
10. An oil tank measures 2.5 m by 3 m by 1.5 m. Find how much oil it contains when full. (Answer in litres.) (1 m^3 = 1000 ℓ)

F

Mixed Examples:

1. Find the perimeter of a room 8.2 m by 6 m.
2. Find the area of a square whose perimeter is 100 ft.
3. A circle has radius 4 ft. Find its area ($\pi = 3.14$).
4. Find the circumference of a circle whose area is 1256 cm^2 ($\pi = 3.14$).
5. Find the area of a circle whose circumference is 880 m $\left(\pi = \frac{22}{7}\right)$.
6. Find the volume of a box measuring 12 cm by 15 cm by 20 cm.
7. A paddling pool measures 30 m by 20 m and is 10 cm deep. How many litres of water does it hold? (1 m^3 = 1000 ℓ)
8. A circular tank has a radius of 30 cm. Find the area of the base ($\pi = 3.14$).
9. A rectangular stamp has an area of 12 cm^2. Its length is $3\frac{1}{2}$ cm. Find the perimeter. (Work in fractions.)
10. A flower bed in a park has a radius of 10 m. The gardener increases the radius by 1 m. Find the percentage increase in area ($\pi = 3.14$).
11. Find the circumference of a circle with a diameter of 10.5 cm. $\left(\text{Take } \pi = 3\frac{1}{7}\right)$ (RSA)

12. Find the perimeter of a square which has an area of 81 cm^2. (RSA)
13. Calculate in cm^3 the volume of a rectangular box measuring 10 cm by 8 cm by 6.5 cm. (RSA)
14. A square has a perimeter of 160 ft. What is its area (in square yards) to one decimal place? (RSA)
15. Find the circumference of a circle with a radius of 35 cm. $\left(\pi = \frac{22}{7}\right)$ (RSA)
16. A square patio has a perimeter of 60 ft. What is its area in square yards? (RSA)
17. Find the perimeter of a rectangular swimming pool which measures 25 m by 10 m. (RSA)
18. What is the circumference of a circle of radius 28 cm? $\left(c = 2\pi r, \text{ take } \pi = 3\frac{1}{7}\right)$ (RSA)
19. Posts costing £1.10 each are placed at one metre intervals round a garden. The garden measures 26 metres by 17 metres. Find the total cost of the posts. (RSA)
20. Find the area of a square which has a perimeter of 60 m. (RSA)

UNIT TWELVE

A Change to 24 hour clock:

1. 3.15 a.m.
2. 7.25 p.m.
3. 8.24 p.m.
4. 6.13 a.m.
5. 4 minutes after midnight.

Change to 12 hour clock:

6. 0215
7. 0928
8. 2112
9. 2319
10. 1255

B How long is it from:

1. 0214 to 0428
2. 1720 to 1815
3. 2124 to 2316
4. 1914 to 2035
5. 8.40 a.m. to 5.30 p.m.
6. 10.30 a.m. to 8.15 p.m.
7. 12.15 a.m. to 8.25 p.m.
8. 12.35 p.m. to 5.30 p.m.
9. 0840 Monday to 0917 Tuesday
10. 9.50 p.m. Saturday to 1.15 a.m. Sunday.

C Use the railway timetable on p. 157 to help you answer the questions.

1. What time does the 0728 from Bournemouth get to London? How long does it take?
2. How long does the 0823 from Brockenhurst take to get to Southampton?
3. How many stations does the 0755 from Bournemouth stop at before it gets to Southampton?
4. I want to be at Southampton Airport by 9 a.m. What train must I catch from New Milton?
5. I catch the 0758 from Lymington Pier. At what time should I arrive at Southampton? How long does it take?

Find how long the following journeys take:

6. The 0919 from Basingstoke to Woking.
7. The 0828 from Winchester to London.
8. The 0906 from Christchurch to Heathrow.
9. The 0640 from Yarmouth to Lymington, then train from Lymington to Winchester.
10. The 0921 from Brockenhurst to Farnborough. Where do you change trains?

D

1. A car travels at 50 km/h. How long does it take to travel 150 km?
2. A man walks 20 miles in 4 hours. Find his average speed.
3. A cyclist can travel at 20 km/h. How long will it take him to travel 30 km?
4. A girl runs 100 m in 35 secs. Find her average speed in m/s. (Answer to the nearest whole number.)

Table 158 (Revised Service) **Mondays to Fridays**
Throughout

Bournemouth to Southampton, Basingstoke and London

			J			G	A						B		E	J			A	D				J	
Bournemouth d		07 10	07 28							07 55				08 12		08 41					09 00		09 12	09 41	
Pokesdown d		07 13	07 32							07 59				00 15									09 15		
Christchurch d	07 17		07 35							08 03				08 19							09 06		09 19		
Hinton Admiral d		07 22								08 08				08 24									09 24		
New Milton d		07 26	07 42							08 13				08 29							09 13		09 28		
Sway d		07 31								08 18				08 35									09 33		
Brockenhurst a		07 36	07 49							08 22				08 39							09 20		09 38		
Yarmouth Slipway Ship d			*06 40*					*07 10*													09a05				
160 Lymington Pier d			*07 28*					07 58													*08 58*				
160 Lymington Town d			*07 30*					08 00													*09 00*				
Brockenhurst d		07 37	07 49					08e13	07 56	08 23				08 41							09 21		09 39		
Beaulieu Road d		07 43												08 47									09 45		
Lyndhurst Road d		07 47						08 22	08 05					08 51									09 49		
Totton d		07 51						08 27	08 10					08 55									09 53		
Redbridge 165 d		07 53						08 29	08 12					08 58									09 55		
Millbrook (Hants) d		07 57						08 33	08 16		←			09 01				←					09 59		
Southampton 165 a		07 59	08 04					08 36	08 19	08 38	08 36			09 04		09 09		09 04			09 35		10 01	10 09	
Southampton 165 d		07 59	08 05					08 42	08 28	08 39	08 42		08 56	09 13		09 09		09 13			09 37		10 13	10 09	
St. Denys 165 d		03 05						→	08 38		08 47			→				09 17					→		
Swaythling d		03 08							08 42		08 50							09 21							
Southampton Airport d		08 10			←					08 48								09 23			09 46				
Eastleigh d		08c17			08c17		08 43		08a45	08 52	08a54		09 07					09 26	09 43		09 52				
Shawford d		→			08 23		08 49											09 31							
Winchester d			08 21		08 28		08 54			09 01								09 36	09 53		10 01				
Micheldever d	←				08 37													09 45							
Basingstoke d	08o27				08c59	08 52	09a15			09 19					09 25		09 26	09 56	10a15	10 16	10 19				10 26
Hook d	03 33				09 05												09 32	10 02							10 32
Winchfield d	08 37				09 09												09 36	10 06							10 36
Fleet d	08 42				09 14												09 44	10 14							10 44
Farnborough (Main) d	08 46				09 18												09 48	10 18							10 48
Brookwood 156 d	08 53				09 25												09 55	10 25							10 55
Woking 156 a	08 59				09 31	09 15				09 39					09 48		10 01	10 31		10 34	10 39				11 01
406 Heathrow Airport a	*09 50*									*10 50*							*10 50*				*11 50*				*11 50*
Surbiton 152, 156 a	09 13				09 45	*09 45*				*10 08*							10 15	10 45			*11 08*				11 15
London W'loo 152, 156 a	09 29		09 19		10 01	09 41				10 04					10 16	10 18	10 31	11 01		11 03	11 05			11 18	11 31

(British Railways Board)

5. A motorist travels 300 miles in 8 hours. What is his average speed?
6. A robot travels at 10 km/h. How far will it travel in 4 hours?
7. A train travels at 120 m.p.h. How far will it travel in $2\frac{1}{2}$ hours?
8. A train leaves at 0810 and arrives at 0940. It has travelled 60 miles. Find its average speed.
9. A plane travels at 600 m.p.h. It travels a distance of 150 miles. How many minutes does the journey take?
10. A man travels 10 miles to the station in his car at an average speed of 20 m.p.h. Then he travels a further 175 miles on the train at an average speed of 70 m.p.h. How long does the total journey take?

E

1. Change 60 km/h to metres per second.
2. Change 400 metres per second to km/h.
3. Change 80 metres per minute to centimetres per second.
4. Change 88 feet per second to miles per hour.
5. A boy notices the train he is on travels the 3 miles between stations in 3 minutes. What is the average speed of the train?
6. A car travels at 30 m.p.h. How far can it travel in 6 minutes?
7. An athlete wishes to run 2000 m in 6 minutes. What speed must he run at? (Answer in kilometres per hour.)
8. An aeroplane travelled 4000 miles in 6 hours 40 minutes. How fast was it travelling?
9. A man rows at 5 km/h along the river. How long will a journey of 8 km take?
10. A rocket travels at 1400 km/h. How far will it travel in 1 second? (Answer to 2 sig. fig.)

F

1. A woman is paid £2.25 per hour. How much will she earn for a 16-hour week?

2. A builder is paid £3.50 per hour and double pay on Sundays. He works 42 hours including 2 hours on Sunday. How much does he earn?

Use this table for the following questions:

	Factory wages	*36-hour basic week*	
Grade	*Basic rate*	*Overtime at time and a half*	*Sundays at double time*
A	£1.50	£2.25	£3.00
B	£2.00		
C	£3.00		

3. Copy and complete the above table.
Use your table to calculate the following people's wages.
4. Jill is on grade A. She works 36 hours.
5. Mary is on grade B. She works 50 hours, but does not work on Sunday.
6. Nick is on grade B. He works 42 hours, including 2 hours on Sunday.
7. George is on grade C. He works 45 hours including 5 hours on Sunday.
8. Simon is on grade B. He works $38\frac{1}{2}$ hours, but does not work on Sunday.
9. Karen is on grade C. She works $42\frac{1}{2}$ hours including $2\frac{1}{2}$ hours on Sunday.
10. Andy works 9 a.m. to 6 p.m. every day with one hour break for lunch. He does not work on Sunday.

G Mixed Examples:

1. Change 2.07 p.m. to the 24 hour clock.
2. Change 1759 to the 12 hour clock.
3. A man is driving to the station hoping to catch the 1723 train. He hears on his car radio that it is now a quarter to five. How long has he got to catch the train?
4. A train left Newcastle at 1412 and arrived in London at 1736. The train was 5 minutes late. How long should the journey have taken?
5. A television programme started at 8.55 p.m. and ended at 10.15 p.m. How many minutes did it last?
6. A car travelled 150 miles in $4\frac{1}{2}$ hours. Find its average speed.
7. It is $1\frac{1}{2}$ miles to the station. A man has 25 minutes to catch the train. How fast must he run?
8. A man is paid £2.40 per hour. He works $7\frac{1}{2}$ hours per day, 5 days per week. How much does he earn per week?
9. A bank clerk is paid £14 200 per year. How much is she paid per month? (Answer to the nearest £10.)
10. A car travels at 44 m.p.h. for $2\frac{3}{4}$ hours. How far has it travelled?
11. A car journey between Leeds and Keswick is 96 miles and the journey takes $3\frac{3}{4}$ hours. What is the average speed in miles per hour for the journey? (RSA)
12. A train left Birmingham New Street at 1820 and arrived in London Euston at 2007. If the train was running 14 minutes late, how long should the journey have taken if the train had run to schedule? (RSA)
13. A jogger runs 35 miles in a week. How far does he run in 52 weeks? (RSA)
14. If a clock gains 75 seconds a day, how many minutes will it gain in 8 days? (RSA)
15. A person is paid £2.00 an hour for a basic 35 hour week, with overtime at time and a half. What would be the person's pay for a 41 hour week? (RSA)
16. Alan Jones makes parts for cars and averages 140 units a day over a 5 day week. What will be his weekly earnings if he is paid 15 p per unit? (RSA)
17. The time in Mexico City is 6 hours behind London time. What is the time in Mexico City when it is 2 p.m. in London? (RSA)
18. A flight to Italy was due to take off at 1820 hours. It actually took off at 2012 hours. By how long was it delayed? (RSA)

19. A part time clerk averages 12 hours per week during a month of 4 weeks. What will be her average wage for a month at £1.75 per hour?

20. A lady has been investing £5 in Premium Bonds every month, starting in July 1979. How much will she have invested in bonds by the end of 1982? (RSA)

UNIT THIRTEEN

These questions are all taken from old RSA papers. You should try to complete each exercise in half an hour. Don't forget to check your answers.

A

1. £1.42 × 5
2. 100 × 2.19
3. 477.6 ÷ 1000
4. What is 15% of £4.50?
5. A clock gains 5 seconds every day. How many minutes will it gain in 24 days?
6. On a journey of 256 miles a car averages 32 miles per gallon. If the petrol costs £1.72 per gallon, how much will the trip cost?
7. A train left Bradford at 0847 and arrived in London at 1131 having been delayed in Doncaster for 12 min. How long should the journey have taken if there had been no delay?
8. A man is paid £2.40 per hour for a 40-hour week. For overtime he is paid an extra £1 per hour. How much will he earn in a week when he has worked 56 hours?
9. Thirty-six marks were scored out of a possible total of 45 marks in a general knowledge quiz. Calculate this mark as a percentage of the total.
10. The entrance cost to a museum was 30 p per person. If takings for a particular day were £153.90, how many people paid to enter the museum?

B

1. 7214 ÷ 100
2. £36.15 × 5
3. Find the cost of five postage stamps at 17 p each and six stamps at 13 p each.
4. What is 12% of £15?
5. A cyclist travels 14 km in 35 min. What is the average speed in kilometres per hour?
6. Express $\frac{5}{8}$ as a percentage.
7. A train left London at 1410 hours and arrived at its destination at 2135 hours. How long did the journey take?
8. A train completed a journey of 273 miles at an average speed of 65 miles per hour. How long did the journey take? (Answer in hours and minutes.)
9. A schoolboy wants to earn £33 to buy a second hand cycle. He decides to wash cars in his neighbourhood. How many must he wash to raise the money if he charges 60 p per car?
10. Brian Marshall pays his garage bill of £180 by using his Access credit card and then decides to pay £40 per month off the debt. The rate of interest is $2\frac{1}{2}$% per month after the first payment. How much does he owe *before* he makes his payment at the end of the third month?

C

1. £3.46 × 9
2. £4.90 ÷ 7
3. What is 15% of £4?
4. Find the average of 13, 24, 35, 31 and 42.
5. £73.20 is changed into ten pence pieces. How many are there?
6. Calculate the Simple Interest on £650 invested for 5 years at 8% per year.
7. A man earns £450 per month. He pays 6% of this to a pension fund. How much is left each month?
8. A bicycle is advertised at £81, or one third deposit and 52 weekly payments of £1.25. What is the cost of the bicycle using these Hire Purchase terms?

9. Mr. Jones lives in a council house for which the annual rent is £650. He must also pay annual rates of £130. How much is his combined weekly bill? (Take 1 year = 52 weeks.)
10. This year the rates on a property in Bathgate amounted to £188.80 and this was 18% higher than last year. How much did it amount to last year?

D

1. £18.30 is changed into five pence pieces. How many are there?
2. Find the average of 7, 16, 17, 19, 21 and 22.
3. Victor Speed buys 11 ℓ of petrol each week and drives 19.5 km per litre. How many kilometres does he travel in a year?
4. How many 150 g packs of sweets can be made up from 33 kg?
5. Convert £26 into francs. (£1 = 10.90 francs)
6. Express $\frac{4}{5}$ as a percentage.
7. A dealer sells four cars for £2550, £3240, £5050 and £1820 respectively. What average price does he receive?
8. If the distance between Shrewsbury and London is 150 miles and the second class return rail fare is £27, what is the cost in pence per mile from Shrewsbury to London?
9. The cost of a ticket for a night club in Paris was 236.00 francs. Calculate the cost in pounds if £1 = 11.80 francs.
10. The motor insurance premium for John Smith's car is £184. If he is allowed a 30% No Claims Bonus how much will he have to pay?

UNIT FOURTEEN

A

Do these calculations on your calculator:

1. 5.21×7.46
2. $19.8 - 5.7412$
3. $17.8 + 29.735$
4. $85 \div 77$ (3 decimal places)
5. $(9.86)^2$
6. £5.93 × 18
7. £2.97 ÷ 15 (Answer to the nearest penny.)
8. 1.582 kg × 1.56 (Answer to the nearest gram.)
9. $\sqrt{0.012\,47}$ (3 decimal places)
10. $92.65 \times 8.3 \times 9.2$

B

Now try these on your calculator. Correct all the answers to 4 significant figures:

1. $\frac{7.25 \times 6.3}{8.4}$
2. $\frac{32}{11 \times 7.4}$
3. Change $\frac{4}{9}$ to a percentage.
4. $\frac{1}{2.1}$
5. $7.25 + 6.4 - 8.32$
6. $\sqrt{9.62 + 18.7}$
7. $\sqrt{3.4^2 + 2.1^2}$
8. $3.14 \times (8.6)^2$
9. $\frac{100 \times 27.6}{500 \times 2.75}$
10. Change 2 h 53 m to hours.

C

1. How many pieces of cotton 2.35 m long can be cut from a 500 m roll?
2. Find the area of a room measuring 5.23 m by 8.79 m. (2 decimal places)
3. A man cycles 755 m at 20 km/h. How long does he take? (Answer in minutes and seconds.)
4. A man travels 97 miles by train in 3 h 23 min. Find his average speed. (Answer to 3 sig. fig.)

Questions **5–10** are about this room.

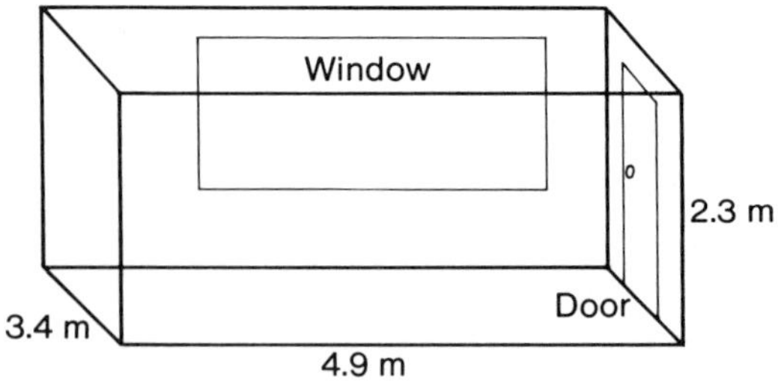

The door measures 2 m by 1 m.
The window measures 2 m by 1.5 m.

5. Find the area of the floor.

6. The carpet is 4 m wide. What length of carpet would you need to buy to cover the floor? Find the cost if the carpet costs £6.99 per square metre.
7. Find the total area of the walls ignoring the window and door.
8. The walls are to be painted, (but not the window and door). Find the area to be painted. A tin of paint covers 25 m^2. How many tins are needed for two coats of paint?
9. Find the cost of the paint at £4.99 per tin.
10. How many lengths of coving each 2 m long are needed to go all round the edge of the ceiling?

D

1. Add £3.17 + £5.83 + £19.16 + 48 p
2. £19.45 × 24
3. £981 − £114.96
4. £6.92 × 3 + £4.62 × 8
5. £84.79 ÷ 13 (Answer to the nearest penny.)
6. Find the total cost of 144 bottles of cola at 22 p each.
7. Find the total cost of 916 components at 6.28 p each. (Answer to the nearest penny.)
8. A man works at $37\frac{1}{2}$ hour week and is paid £2.28 per hour. Find his weekly wage.
9. A man is paid £95.40 for a basic 36 hour week and time and a half for overtime. How much will he be paid for a week in which he works 40 hours?
10. A bill in a restaurant costs £10.84 + 15% VAT. Find the total cost including VAT. (Answer to the nearest penny.)

E

1. If 12 pens cost £3.48, what will 17 pens cost?
2. Two men buy a shop between them in the ratio 3 : 2. The profits for the first year are £1256. How much should each receive?
3. A car travels 8.6 km on one litre of petrol. How far can it travel on 17 litres?
4. A man earns £72.60 for 30 hours' work. How much will he earn for 19 hours at the same rate?
5. Share £12.92 in the ratio 7 : 4. (Answer to the nearest penny.)
6. Some money is divided in the ratio 3 : 4. The larger share is £92. Find the sum of money.
7. Change 214 francs to pounds. (£1 = 10.9 francs)
8. Change £316 to dollars. (£1 = \$1.09)
9. A taxi driver charges £1.96 for 3.5 miles. How much will he charge for 4.6 miles at the same rate?
10. A man is paid a mileage allowance of 18.3 p per mile. How much is he paid in a month if he travels on 23 days and averages 67 miles per day?

F

1. A man travels 118 miles at an average speed of 37 m.p.h. Find how long the journey took?
2. A man travels a journey of 63 miles in 2 h 43 m. Find his average speed.
3. A girl cycled 24 miles in 2 h 53 m. Find her average speed.
4. A boy can jog at 6.2 m.p.h. How far does he expect to go in $2\frac{1}{2}$ hours?
5. A car travels at 93 km in 1 h 12 m. Find its average speed.
6. The distance from Calais to Paris is 315 km. If I leave Calais at 1124 and average 60 km/h, what time should I arrive in Paris?

Questions 7–10 are related.

7. The distance from Plymouth to Santander in Spain is 700 miles. The average time for the ferry crossing is 24 hours. Find the average speed of the ferry.
8. The fare for an adult is £42. Find the cost per mile.
9. Santander is 249 miles from Madrid. If I leave Santander at 0915 and travel at an average speed of 40 m.p.h., at what time will I arrive in Madrid?
10. If I left Plymouth at 0800 the previous day, find the average speed of my entire journey to Madrid.

G Exam Questions:

1. A motorist left London at 9.00 a.m. to travel 196 miles to a destination in Leeds. He covered the first 84 miles at an average speed of 56 miles per hour and the remaining distance at 64 miles per hour. He completed the return journey from Leeds to London in 2 hours 48 minutes.
   a) What time did he reach his destination in Leeds?
   b) What was his average speed for the journey from London to Leeds (give your answer correct to 1 decimal place).
   c) Find his average speed for the return journey. (RSA)

2. Mr. Plant, a groundsman, was responsible for a cricket field and he decided to mix his own grass fertiliser. The field was 130 yards long and 84 yards wide and he spread the lawn food at the rate of 2 ounces per square yard. The lawn food consists of 2 parts bonemeal, 1 part sulphate of potash and 2 parts superphosphate. The bonemeal costs 50p per lb, sulphate of potash costs 60p per lb and superphosphate costs 70p per lb.
   a) Find the weight of lawn food in lbs required for this field.
   b) Find how many lbs of each of the above ingredients are required to make this mixture.
   c) What is the total cost of this mixture? (RSA)

3. In 4 successive months a sales representative travelled 2162 miles, 1826 miles, 1282 miles and 2506 miles respectively. His car used petrol at the rate of 24 miles per gallon and the cost of petrol was £1.82 per gallon.
   a) Find the total cost of the petrol used in the 4 months.
   b) If the petrol could only be purchased in litres find how many litres would be used in the 4 months if 1 litre = 0.22 gallons giving your answer to the nearest whole litre.
   c) If the sales representative had obtained a more economical car which used petrol at the rate of 36 miles per gallon, find how much money would have been saved in the 4 months. (RSA)

4. The table below shows the cost of holidays in Malta.
   a) Work out the cost of a 14-night holiday at the Imperial Hotel for a family of five (two adults and three children). The departure date is July 2nd.

COST PER PERSON IN £s

	HOTEL	PLEVNA		TIGNE COURT		IMPERIAL		FORTINA		MARINA	
	No. OF NIGHTS	7	14	7	14	7	14	7	14	7	14
DEPARTURES ON OR BETWEEN	JUNE 14–27	274	378	249	334	264	361	283	387	247	323
	JUNE 28–JULY 18	279	383	263	350	273	372	281	395	252	328
	JULY 19–AUGUST 22	289	393	273	360	283	382	291	405	262	338
	AUGUST 23–SEPTEMBER 5	279	383	263	350	273	372	281	395	252	328
	CHILD REDUCTION %	30	30	30	30	30	30	30	30	30	30

b) How much more would it cost a family of 4 (2 adults and 2 children) to stay at the Fortina rather than the Marina for 14 nights. The departure date is July 26th.

c) Due to poor bookings, the company decides to reduce all its prices by 7½%. How much would a family of three save (2 adults, 1 child) when staying at the Plevna for 7 nights, departing on July 21st? (RSA)

5. When a petrol station changes from selling its petrol in gallons to litres, each pump has a sticker on it saying:

5 Litres = 1.1 Gallons

a) How many gallons of petrol have been placed in a car when the pump reads 64 litres? Give your answer to 2 decimal places.

b) How many litres equal 1 gallon? (Give your answer to 2 decimal places).

c) If the price of petrol is 40.2 p per litre, how much is this per gallon to the nearest penny?

d) A car travels a distance of 256 miles and uses 38 litres of petrol. Work out the average petrol consumption of the car in miles per gallon. (Give your answer to 2 decimal places.) (RSA)

UNIT FIFTEEN

A Give your answers to 3 significant figures:

1. Change $\frac{17}{23}$ to a percentage.
2. Change $1\frac{24}{83}$ to a percentage.
3. Find $8\frac{1}{2}$ of £298.
4. Find $12\frac{1}{2}$ of 714.
5. On a foggy day 13 girls out of 29 are late for school. Express this as a percentage.
6. Increase £86 by 8%.
7. Decrease £72 by 5%.
8. The price of a coach ticket increases from £5.26 to £6.15. Express this increase as a percentage.
9. Find 9.3% of £217.
10. The volume of oil in a tank decreases from 1862 ℓ to 1514 ℓ. Find the percentage used.

B Give your answers to 2 decimal places:

1. A man buys a painting for £45 and sells it for £60. Find his percentage profit.
2. I bought a car for £4250 and after two years sold it for £2999. Find my percentage loss.
3. A dealer buys a washing machine for £190 and wants to make 24% profit. How much must he sell it for?
4. A coat is priced at £27. In the sale it is reduced by 10%. Find the sale price.
5. A book costs £9.40. It has gone up by 10% from last year. What did it cost last year?
6. A tape recorder costs £45. It is reduced by 15% in a sale. Find the sale price.
7. A girl buys ten blank tapes for £9.95. She sells them to her friends for £1.30 each. Find her percentage profit.
8. A shopkeeper buys 36 kg of butter for £67.20 and sells it at 56 p for 250 g. Find his percentage profit.
9. A dealer sells a car for £2800 having made a 15% profit. What did he pay for the car? (Answer to nearest ten pounds.)
10. A washing machine can be bought for £290 cash or 12 monthly payments of £25.30. Find the extra cost of buying the machine on HP and express this as a percentage of the cost price.

C Give your answers to the nearest penny:

1. A meal costs £8 plus 15% VAT. Find the total cost.
2. A plumber charges £4.50 per hour plus VAT at 15%. It takes the plumber four hours to mend my boiler. How much do I have to pay?

3. A saleswoman pays £23.40 for petrol which includes VAT at 15%. Her company claim back the VAT. How much do they claim back?
4. A painter offers $2\frac{1}{2}$% discount if bills are paid promptly. My bill is £93.70. How much will I have to pay if I pay immediately?
5. This year's rates are £180. They have gone up 8% from last year. How much were last year's rates?
6. A tape recorder costs £34.20 including VAT at 15%. Find the price before VAT was added.
7. A company make a 3% surcharge on holidays. A family booked for two adults and one child. If the original price was £204 for adults and £170 for children, find how much extra the family will have to pay.
8. A boy scored 39 out of 50 in a History exam. "I have done 50% better than my mark last year", he says to his father. What did he score last year?
9. A store offers its staff 15% discount. How much will a member of staff pay for a clock priced at £29.90?
10. Goods in a sale have 20% off. I paid £9.50 for some earrings in the sale. What did they cost before the sale?

D

A restaurant adds 15% to the bill for VAT and then 10% for service:

1. Find the cost of a meal priced at £7.50.
2. Find the cost for two people having meals costing £5.50 and £6.00 and a bottle of wine between them costing £3.50.
3. Two men are given a £23.00 total bill. They think the service was terrible and refuse to pay for it. How much do they pay?
4. Another man pays £14 for his meal and VAT but not service. How much VAT has he paid?
5. A family have a total bill of £34.76. Find the price before VAT and service were added.
6. A cooker is priced at £240. The shop offers 10% reduction for cash and then a further $2\frac{1}{2}$% reduction for quick payment. Find the price the customer actually pays if he gets both these discounts?
7. A salesman is paid $17\frac{1}{2}$% of the value of the goods he sells. He pays 30% tax on his earnings. If he sells £18 000 worth of goods in a year, how much will he receive?
8. A shopkeeper buys a dress for £13. She adds 30% for her profit. In the sale she reduces the dress by 20% of the marked price, and sells it. What percentage profit does she actually make on her outlay?
9. An antique dealer buys a chair for £25. He restores it and prices it at £70. He reduces it by 50% in his sale and then sells it. What percentage profit does he actually make?
10. A girl scores 40% in a 'mock' examination. In the actual exam she does 10% better but still fails. She resits the exam and gets 20% more than the previous time. Find her final mark. (Answer to the nearest per cent.)

E

Exam Questions:

1. A family of two adults and two children took a 14 day holiday in a hotel in Scotland. The hotel charges £8.50 per day for each adult and £6 per day for each child. During the holiday the family travelled a total distance of 1650 km. The car used petrol at the rate of one litre every 6 km and costs 38p per litre.
   a) Find the cost of the hotel for this family if VAT at 15% is added to the total cost.
   b) What were the total petrol costs for the holiday?
   c) Express the total petrol cost as a % of the total cost of the holiday giving your answer to the nearest whole number.

(RSA)

2. A woman decides to buy a cleanomatic washing machine on hire purchase and finds the machine available at three different shops offering the following special deals:

SHOP A	DEAL
CASH PRICE £302.60	NO DEPOSIT. 12 MONTHLY PAYMENTS OF £31.68

SHOP B	
CASH PRICE £299.50	⅕ DEPOSIT. 24 MONTHLY PAYMENTS OF £13.39

SHOP C	
CASH PRICE £285.21	33⅓% DEPOSIT. 18 MONTHLY PAYMENTS OF £15.47

a) What is the total cost of each deal?

b) Work out the interest payable on each deal and express it as a percentage of that particular cash price. (Give your answer to 2 decimal places.)

c) Shop B makes a profit of 17½% on cost when selling its machines at the cash price. How much do they pay for a washing machine when buying it for the shop? (Give your answer to the nearest penny.) (RSA)

3. Mr. Brown, a builder, bought a piece of land for £93000 and split it up into 30 equal plots. He then built 12 detached bungalows and 18 detached houses. It cost him £22500 to build each bungalow and £29750 to build each house. He sold each bungalow for £31750 and each house for £38500.

a) Work out his total profit when all the properties are sold. (The building costs do not include the price paid for land).

b) During the year, Mr. Brown made a profit of £1.5 million. Express the profit made on his site as a percentage of his total yearly profit. (Answer to 1 decimal place.)

c) How much was paid in stamp duty to the government when all the properties had been sold? Stamp duty was payable on the purchase price of each property sold. The rates payable are given below:

Purchase Price of Property	*Percentage Payable*
£25001–£30000	1%
£30001–£35000	1½%
£35000 upwards	3%

(RSA)

4. Mrs. Jakes made enquiries about the purchase of a Hi-Fi unit as a birthday gift for her husband. The Hi-Fi unit had a marked price of £480 and she found that it could be bought on Hire Purchase by paying a deposit of 15% and 24 monthly payments of £18 or by cash when a discount of 12½% is allowed off the marked price.

a) Find how much Mrs. Jakes would pay if she decided to buy the Hi-Fi on Hire Purchase.

b) How much would it cost her if she paid cash?

c) Find the difference between the total Hire Purchase and the cash price and express this difference as a percentage of the marked price. (RSA)

5. The total cost of running a car for one year is £1952 and this amount can be split into three sections:

(i) Tax, Insurance and Depreciation;
(ii) Repairs, Service and Tyres;
(iii) Petrol;

in the ratio 5 : 4 : 7.

a) Work out the amount spent by the car owner in each of the three sections.

b) If the repairs cost £241.68 and a set of new tyres was £65.48, how much was spent servicing the car during the year?

c) If the owner paid, on average, £1.84 per gallon for his petrol during the year, how many gallons did he buy? (Give your answer to the nearest whole gallon.)

d) If the car was sold at the end of the year at a loss of 28%. How much did the car cost when new if the loss was £980? (RSA)

UNIT SIXTEEN

A Give your answers to the nearest penny:

1. A man invests £500 for 3 years at 8% Simple Interest. How much interest does he receive?
2. A woman invests £280 for $2\frac{1}{2}$ years at 7% Simple Interest. How much does she have altogether after $2\frac{1}{2}$ years?
3. A boy is given £500. He invests it at $11\frac{1}{2}$% Simple Interest for 5 years. How much interest does he receive?
4. A girl wins £1000. She spends £242 on a television and saves the remainder, which she invests at $9\frac{1}{2}$% Simple Interest. How much does she have after one year and three months?
5. A pensioner invests his money in an Investment Account giving 12% Simple Interest. He receives the interest monthly and gets £52.12 per month. How much money has he invested?
6. A company borrows £20 000 from a bank at 16% Simple Interest. They pay back £24 242. How long, to the nearest month, did they keep the loan?
7. A bank has a variable rate of interest. I invest £400. I get £13 interest for the first six months, and £13.50 for the next six months. Find the rate of interest for each half yearly period.
8. A man wishes to borrow £8000 to extend his house. The bank gives him a loan of £5000 at 12% Simple Interest and he borrows the rest from a finance company at 19%. After one year he pays back the finance company. After a further 18 months he pays back the bank loan. How much interest does he pay altogether?
9. A building society offers me 8% Simple Interest on an ordinary account and $9\frac{1}{2}$% on a special account. If I draw money from the special account I lose two months' interest. If I have £100 and have to withdraw it after one year which account would give me the most interest and by how much?
10. A boy is given £500. He wishes to invest enough to give him £1 a week interest. If he can get 12% interest how much money can he spend now?

B Find the amount after compound interest has been added:

1. £200 for 2 years at 8%.
2. £500 for 3 years at 10%.
3. £700 for 2 years at 9%.
4. £1600 for 3 years at 12%.
5. £5200 for 2 years at $14\frac{1}{2}$%.

Find the compound interest:

6. £300 for 2 years at 9%.
7. £250 for 3 years at 7%.
8. £900 for 4 years at $6\frac{1}{2}$%.
9. £1500 for 2 yeas at $6\frac{3}{4}$%.
10. £2965 for 3 years at $8\frac{1}{4}$%.

C

1. A factory owner spends £25 000 on a new machine. It depreciates by 10% per year. Find its value after two years.
2. A man sells his business for £30 000 and retires. He spends 10% of his capital the first year and 15% of the remaining money the next year. How much has he left after two years?

3. A couple buy a house for £32 000. During the first year it increases in value by 8%. During each of the next two years it increases by 7%. Find the value of the house after the couple have owned it for three years. (Answer to the nearest pound.)
4. A man bought a car for £5800. After one year he was offered £4700 for it. He did not sell the car. After another year he sold it for £4200. If the prices he was offered for the car were fair what was the percentage depreciation each year?
5. A company buy a computer for £2000. They estimate its value will depreciate by 10% a year for the first two years and then by 20% per year. Find its value after four years. (Answer to the nearest pound.)

D

1. A cleaner is paid £1.90 per hour and time and a half for Saturdays. She works from 7 a.m. to 9 a.m. each morning except Sunday. How much is she paid for a week's work?
2. A waitress is paid £1.60 per hour and double time after midnight. One evening she works from 7 p.m. to 1 a.m. How much does she earn?

A factory pays £2.24 per hour for a 36 hour week and time and a half for overtime:

3. John Smith works 39 hours. How much does he earn in a week?
4. Andrew Jones works $42\frac{1}{2}$ hours. How much does he earn in a week?
5. David Sullivan earns £112.56. How many hours does he work?
6. The wages at the 'Topcar' factory are as follows:

	Basic 30 *hours*	*Overtime time and a half*	*Sundays double time*
Scale I	£2.16		
Scale II		£4.26	
Scale III			£6.56

Copy and complete the table.
Find the wages of the following people:

7. Sam Brown is on scale I. He worked 44 hours, but did not work on Sunday.
8. Carol White is on scale II. She worked 35 hours including 4 on Sunday.
9. Fred Black is on scale III. He earned £159.90 and did not work on Sunday. How many hours did he work?
10. The manager increases all the wages by 4%. How much will they each earn next week if they work the same hours?

E

1. A part-time shop assistant is paid £24.00 basic wage plus 3% of sales. She sells £315 worth of goods. How much does she receive?
2. A saleswoman is paid commission only. She is paid 25% of sales. How much is her average weekly pay if she sells £24 215 worth of goods in a year?
3. A girl packs toys. She gets a basic wage of £1.80 per hour and is expected to pack 80 toys. She is paid 5 p bonus for every extra toy she packs. On a certain day she works $7\frac{1}{2}$ hours and packs 720 toys. Find her total pay for the day.
4. A salesman is paid £180 per month plus 7% commission. If he earns £240 in total find the value of his sales. (Answer to the nearest pound.)
5. A salesman is paid 15% commission on goods sold from stock and 10% on special orders. In a week he sells £218.90 worth of goods and gets £93.80 worth of orders. Find his total commission.
6. A woman sells goods from a catalogue and gets 25% commission. She sells £26.30 worth of goods but spends 54 p on postage. How much profit does she actually make?
7. A small company employs three salespeople, who are paid 5% commission, and two office staff. During the year and three salespeople received £314, £182, and £423 in commis-

sion. At Christmas the company gave a 1% bonus on total sales to be divided equally between the five staff. How much did they each recieve?

8. A representative is paid £3620 per year plus 7% commission. During his first year he sells £18 250 worth of goods. During the second year he increases his sales by 23% and the company increase his basic wage by 8%:
   (a) Find his commission for the first year.
   (b) Find his total salary for the second year.

F For this exercise assume the following tax allowances, which are based on the 1984/5 figures.

Allowances
Single Person £2005
Married Person £3155
Superannuation and National Insurance Contributions Tax Free

Rates

£1 *up* to £14 600	30%
£14 601–£17 200	40%
over £17 200	50%

Find the tax paid by the following people:

1. A single woman earning £3300 and paying £104 National Insurance per year.
2. A married man earning £6850 and paying £219 National Insurance and Superannuation per year.
3. A single woman earning £58.97 per week and paying £1.90 National Insurance. (Give weekly tax.)
4. A married man earning £9148 and paying 6% superannuation and £100 National Insurance per year.
5. A married man earning £9148 and paying 8% superannuation and £120 National Insurance per year.
6. A single woman earning £11 987 with a total of £2675 in allowances.
7. A married man earning £24 900 and paying 8% superannuation and £104 National Insurance.
8. A single woman earning £1614.
9. A married woman earning £22 400 after deductions for superannuation and National Insurance.
10. A single person earning an average of £1900 per week and paying 7% of his earnings to tax free funds.

G Exam Questions:

1. James Robinson bought 2000 (£1) Premium Bonds and in the course of 3 years won prizes totalling £480.
   Celia Isherwood invested £2000 in a Building Society at 11% compound interest for 3 years.
   a) Calculate the amount, correct to the nearest penny, invested in the Building Society after 3 years assuming there had been no withdrawals.
   b) Express the average yearly prize money which James received as a percentage of the amount invested in Premium Bonds.
   c) Which was the better investment and by how much? (RSA)
2. A salesman is paid a salary of £350 a month. In addition he receives a commission of 12% of the total sales. His total sales in one year amounted to £82500.
   a) Calculate the total amount of money earned in that year by the salesman (including his salary and commission).
   b) Find his average net monthly pay assuming deductions of 28 pence in the pound.
   c) If the value of the sales in the following year fell by 10%, calculate the amount received in commission. (RSA)
3. Mr. Able invested £3000 in 2 year Local Authority Bonds which paid 9% compound interest but he has to pay tax at the rate of 30% on this interest. He could have lent this

money to a friend who was willing to pay 6% Simple Interest for the same 2 year period.

a) How much interest would he have received if he had agreed to lend the money to his friend?
b) How much interest did Mr. Able receive from the Local Authority Bonds after the deduction of tax?
c) Find the difference in interest between these two possible investments and express this difference as a percentage of £3000. (Give answer correct to one decimal place.) (RSA)

4. Mr. Small, a married man earns a salary of £7300 per year. After deduction of the following allowances from gross pay, income tax is charged at 30%.

Personal allowance	£2445
Superannuation	£365
Building Society Interest	£630

a) How much tax does Mr. Small pay in one year?
b) Find what percentage of his gross salary he pays in tax giving your answer correct to one decimal place.
c) If Mr. Small receives a pay rise of 10% and his total allowances remain the same find his new annual tax bill. (RSA)

5. Miss Lewis, a shop assistant, is paid a basic wage of £28.50 a week and in addition receives a commission at the rate of 9p in the £ on the value of goods sold. In 1981 she sold £35 620 of goods but in 1982 her sales fell by 15%.

a) Find the total annual wage earned by Miss Lewis in 1981.
b) How much commission did Miss Lewis receive in 1982?
c) Find the average weekly wage earned by Miss Lewis in 1982 if her basic wage remained unchanged and give your answer correct to the nearest penny (one year = 52 weeks). (RSA)

6. A bank clerk earns £7800 per year. He is paid monthly and 30% of his gross salary is taken in deductions.

a) Work out his monthly *net* salary.
b) Each month four deductions are made from his salary:

   INCOME TAX £105.50
   NATIONAL INSURANCE £35.70
   SUPERANNUATION £32.50
   BANK LOAN REPAYMENT?

   Work out how much is deducted from his salary per month for his Bank Loan Repayment and express this as a percentage of his monthly net salary. (Give your answer correct to 2 decimal places).
c) In 1984 the clerk is promoted and receives an extra £2300 per year. A further pay rise of 6.5% of his new salary is awarded to him by the bank. If his deductions are increased to 32% of his gross salary, what will his new *net* monthly salary be? (Give your answer to the nearest penny). (RSA)

UNIT SEVENTEEN

A

1. The rateable value of a house is £300. If the rate levied is 87 p in the pound, find the rates to be paid.
2. A house has a rateable value of £214. The council levy a rate of 104 p in the pound. Find the amount to be paid.
3. A man pays £318.55 a year in rates. The rate levied is 115 p in the pound. Find the rateable value of his house.
4. A couple move to a house with a rateable value of £350. They pay £399 in rates. Find the rate levied.
5. A person pays £282.75 in rates. The rate levied is 87 p in the pound. Find the rateable value of the person's house.

6. The rateable value of a town is £2160 000. The council wish to raise £54 000 to repair the town hall. Find the rate they must levy to pay for it.

7. The rateable value of a town is £450 000. A 3 p rate is raised for the fire service. How much do they receive?

8. The product of a penny rate for a city is £315 000. Find:
   a) The rateable value of the city.
   b) The total amount raised if the rate levied is 85 p in the pound.

9. A house has a rateable value of £395. The rate levied is 135 p in the pound. £23.70 of the rates paid goes to the parish. Find:
   a) The rate paid by the householder.
   b) The rate levied for the parish.

10. Mr. Green lives in a bungalow with a rateable value of £260. In 1984 the rate levied is 114 p. In 1985 the rate levied is 118 p. Mr. Green chooses to pay his rates in ten equal instalments. Find:
   a) The amount paid in each instalment in 1984.
   b) The percentage increase in 1985.

B

1. A man pays £12 a week in rent and £104 per year in rates for his flat. Find his total payments per month. (1 year = 52 weeks = 12 months)
2. A couple bought a house with a rateable value of £290. They have to pay £104.80 a month in mortgage repayments. The rate levied is 132 p in the pound. Find their total monthly cost for the house.
3. A student can afford £15 per week for a flat. He finds a flat with a rateable value of £104. The rate levied is 98 p in the pound. The landlord wants £580 per year rent excluding rates. Can the student afford the flat and by how much?
4. A couple borrow £24 000 from a Building Society at 14% interest to buy a house. During the first year they pay back £500 of the loan plus interest. Their rates are £290 per year. Find the total monthly cost of the house.
5. The rateable value of a house is £190. The rate levied is 112 p in the pound. The owner lets the house in order to make £1000 per year profit. How much does he charge per month including rates?

C

1. Copy and complete this bill:

Tower Gas Board		*Account No.* 123456		*Date* 24/7/84	
Meter readings		*Cubic feet (hundreds)*	*Therms*	*Price per therm*	*Amount*
Present	*Previous*				
6848	6672	....	....	35.2 p	....
Standing charge					£9.90
VAT @ 0%					....
		Amount now due			....
Note: To calculate therms used multiply the difference in meter readings by 1.032					

2. Copy and complete this bill:

Lightning Electricity Board			*Period ending* 4/2/85				
Meter reading		*Units used*	*Pence per unit*	*Amount £*	*Standing charge £*	*VAT code*	*Total £*
Present	*Previous*						
37357	35965	....	4.83	....	6.37	0	....
16756	15433	....	4.55	....	0.00	0	....
		Hire purchase on cooker		33.54		15%	....
				Amount due			....

3. Copy and complete this bill:

Telefon	*Number* 453323		*Date* 9/10/85	
Rental	*from*	*to*	*£ quarterly rate*	*Total*
	1 OCT	31 DEC	16.15	....
Metred units	*date*	*reading*	*units used at* 4.40 p	
	5 JUL	0016135		
	4 OCT	0016863	....	....
Value Added Tax at 15.00%				....
			Total Payable	....

4. Mr. Bell receives a telephone bill. He pays £14.50 rental plus 5 p per unit plus 15% VAT. If he uses 212 units calculate the total bill.
5. If Mr. Bell uses 415 units during the next quarter calculate his next bill.
6. Mr. Therm has just received his gas bill. He pays a standing charge of £9.80 plus 38.5 p per therm. 168 therms were used during the quarter. Calculate the total bill.
7. Mrs. Amps pays 4.7 p per unit for electricity plus a standing charge of £6.25 per quarter. If she uses 97 units of electricity find how much she must pay.
8. A family pay £16.00 standing charge for their telephone. Units cost 5 p and 15% VAT is added to the bill. If the total bill is £26.22 find how many units they used.

D Exam Questions:

1. A householder's electricity meter readings for one quarter were as follows:
   End of quarter 13550 units
   Beginning of quarter 12910 units

   There is a quarterly standing charge of £6.25 and the cost of each unit is 4.85p. Each quarter a further £18.25 is added to the bill for a period of four years as an instalment payment for a cooker bought on hire purchase.
   a) Calculate the total cost of the quarterly bill including the hire purchase instalment.
   b) What will be the total cost of the cooker when all the instalments have been paid?
   c) If the number of units are increased by 15% in the following quarter, calculate the number used. (RSA)

2. A telephone bill is received as below:

YORKSHIRE TELEPHONE COMPANY

Mr. C. Smith
12 Carr Bottom Street
Leeds, West Yorks.
LS4 6ZT

Telephone No.	**Date of Bill**
Leeds 672947	31 Aug 83

Rental	**From**	**To**	**Quarterly Rate**	**£**
	1 Sept	30 Nov	£13.50	13.50

Metered Units

Date	**Meter Reading**	**Units Used**	**£**
24 June	000124		
25 Aug	000367		
Units @ 4.30p		(A)	(B)
Total (exclusive of VAT)			(C)
Value Added Tax at 15%			(D)
		Total Payable	(E)

In all your workings work to the nearest penny.

a) Work out:
   - **A** The units used.
   - **B** The cost of units used.
   - **C** The total bill excluding V.A.T.
   - **D** The V.A.T. payable on **C.**
   - **E** The total bill payable including V.A.T.

b) In the following quarter the V.A.T. payable was £3.96. How many units had been used if the standard charges remain the same? (RSA)

3. A bungalow has a rateable value of £338. In 1981, the water rate was 13.6 p in the pound and the sewerage environmental services rate was 16.2p in the £.
   a) Find the total cost of these services for 1981 giving your answer correct to the nearest penny.
   b) In 1982 the water rate is increased by 1.1p in the £ and the sewerage and environmental rate by 1.4p in the £. Find the total increase in these rates for 1982.
   c) What is the increase in rates as a percentage of the total cost in 1981 giving your answer correct to one decimal place? (RSA)

4. A quarterly account for the use of a telephone was made up as follows:
   Rental £15.60 a quarter
   Metered units at 4.3p per unit
   VAT at 15% of the total bill.
   At the beginning of the quarter the meter reading was 6776 and at the end 7356 units.
   a) What is the total cost of the bill exclusive of VAT?
   b) Find the total cost of the bill inclusive of VAT giving your answer correct to the nearest penny.
   c) If the number of units used in the following quarter are reduced by 15% calculate the number used. (RSA)

5. The General Rates charged for a certain house for 1981/82 and 1982/83 are shown below.

	Rateable Value	*Rate in £ (pence)*	*Total Due for Payment*
1981/82	124	120.60	
1982/83	124	156.90	

The total due for payment is found by multiplying together the rateable value and the rate in £.

   a) (i) work out the total due for payment in 1981/82 and 1982/83, give your answers in pounds and pence. (Write your answers in your answer book.)
      (ii) Express the rise in the 1982/83 bill as a percentage of the 1981/82 bill. (Give your answer correct to two decimal places.)
   b) The Water Rates for the same periods were as follows
      1981/82 £52.96
      1982/83 £60.80
      If a householder paid the bills in half yearly instalments, what would his total rates bill have been for a six month period during
      (i) 1981/82;
      (ii) 1982/83?
      (Total Rate = General Rate + Water Rate.)
   c) For 1983/84 the General Rate is increased by 7½% and the Water Rate is increased by 7¼%. Work out the total rate bill for the period 1983/84. (RSA)

UNIT EIGHTEEN

A Find the length of the unknown side on each triangle:

1.

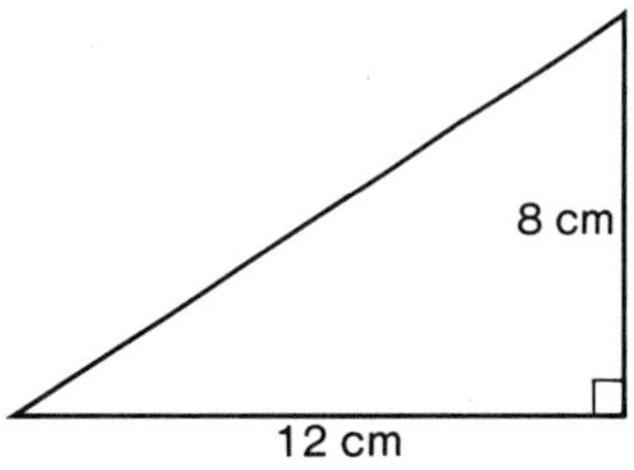

2.

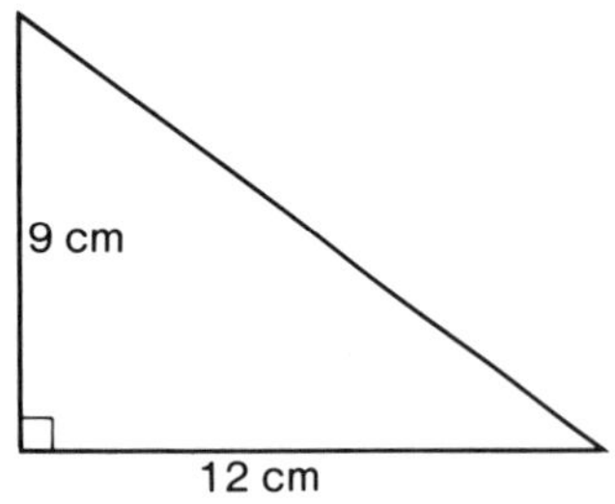

3.

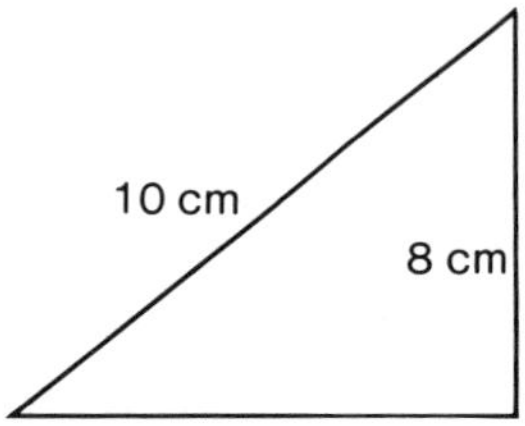

4.

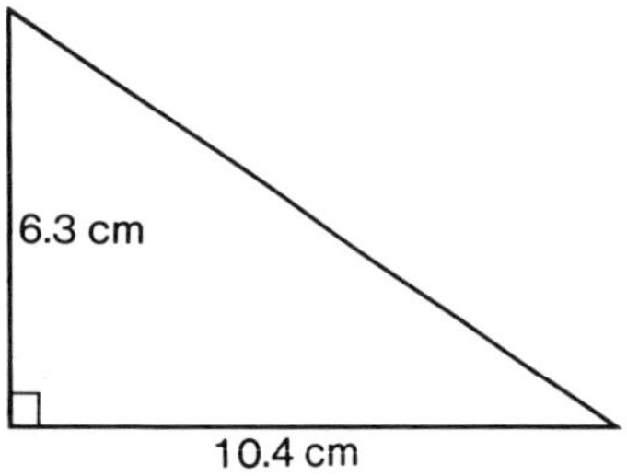

5.

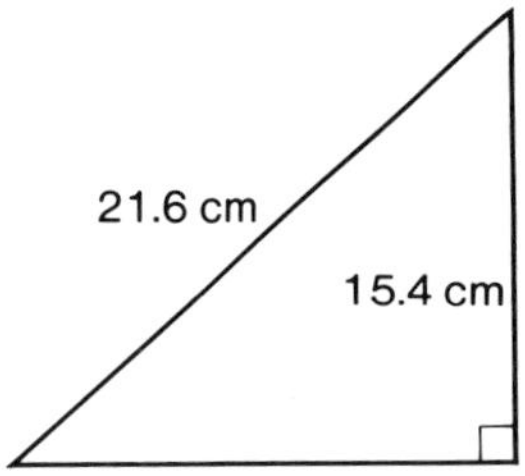

6.

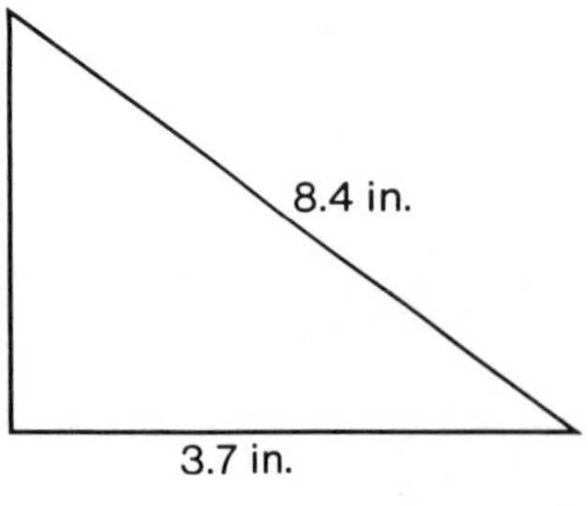

7. A ladder is 7 ft long. If the bottom is placed 4 ft from a wall, how high will the ladder reach?

8. I have a square metre of cloth. I cut it into two equal triangles. Find the perimeter of one of these triangles.

9. A man is in a field with a bull. The bull sees the man and both run towards the gate. The man needs 1 second to get over the gate. If the bull runs at 7 m/s how fast must the man run to escape the bull? (Answer to 2 decimal places.)

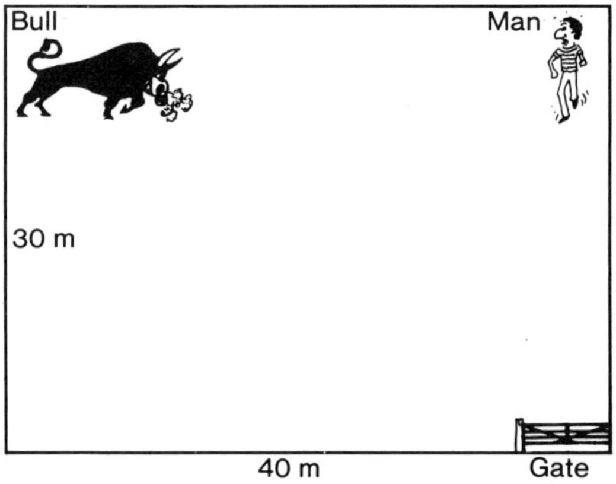

10. Some boys make a flag by sewing four pieces of ribbon on to a handkerchief. If the handkerchief is a 30 cm square, how much ribbon do they use?

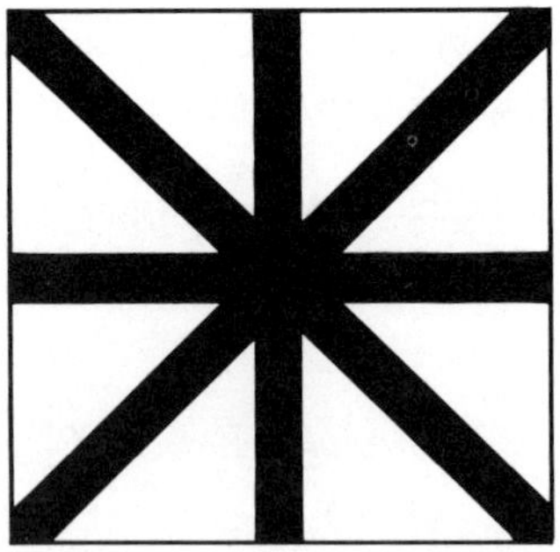

B Give your answers to 3 sig. fig. ($\pi = 3.14$).

1. Find the area of a circle of radius 6.3 m.
2. Find the area of a circle of diameter 23 cm.
3. Find the circumference of a circle of diameter 6 in.

4. Find the area of a circle of circumference 29 cm.

5. Find the circumference of a circle of area 32 m^2.

Find the shaded area on these shapes. ($\pi = 3.14$):

6.

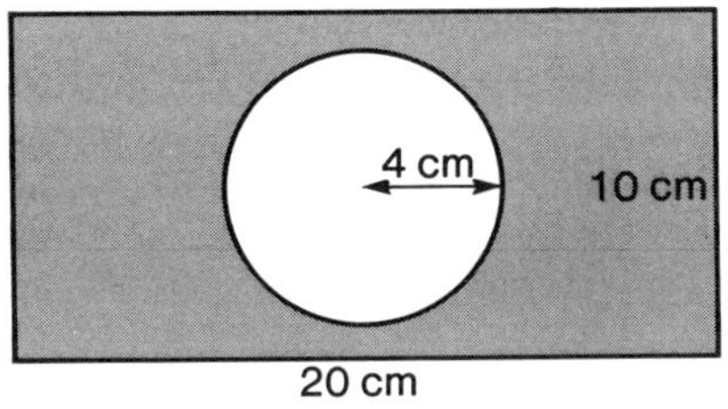

7.

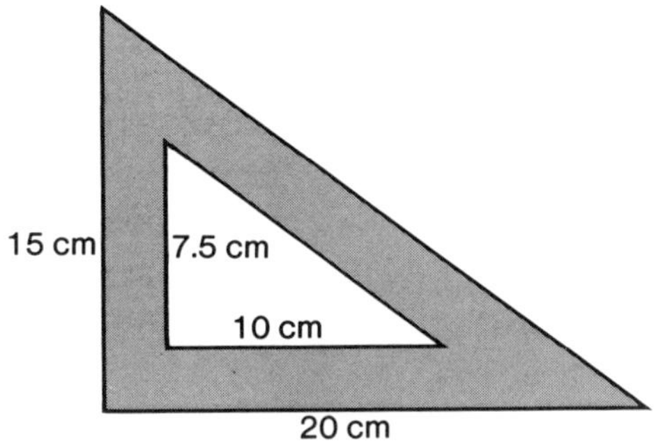

8.

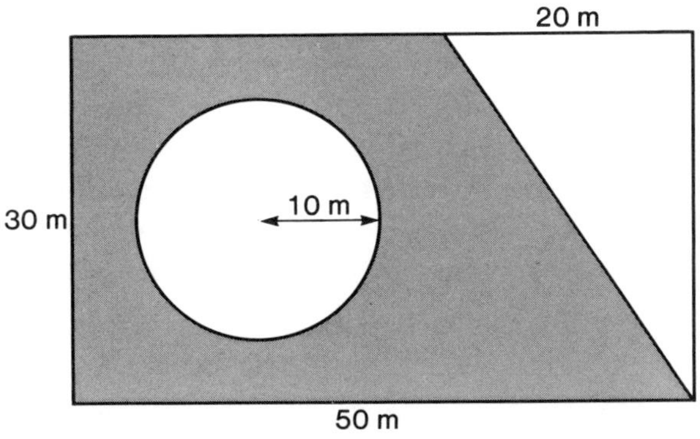

9.

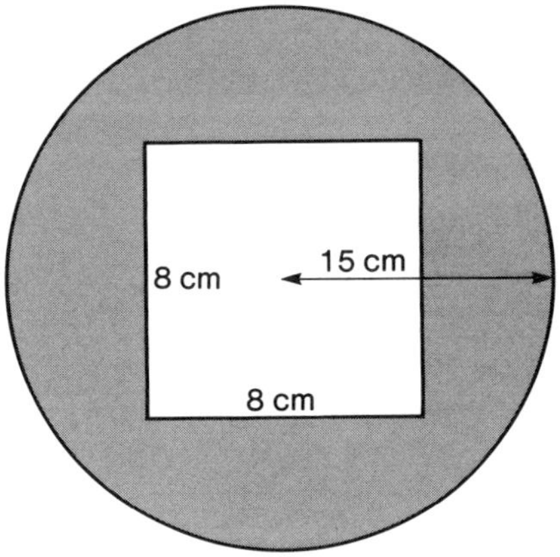

10. A man has a new garden 60 m by 20 m. He draws a sketch of the garden he wishes to make. The vegetable plot is to be one third of the total area of the garden.
   a) Find the area of the vegetable plot.
   b) Find the area of the grass.
   c) If grass seed costs 35 p to cover 10 m^2, find the total cost of grass seed used.

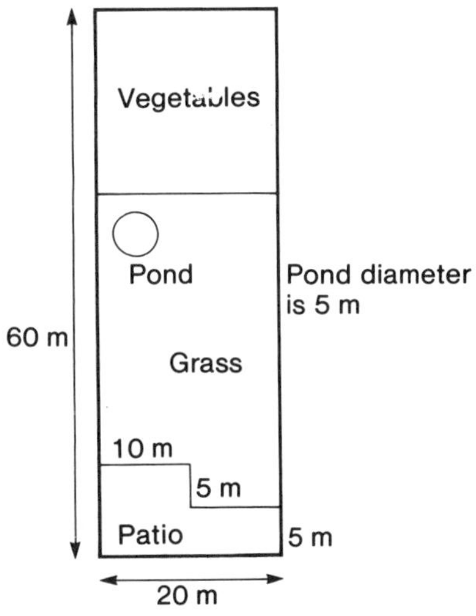

C

1. Find the volume of a rectangular box measuring 2.6 m by 1.9 m by 85 cm.
2. A room has a volume of 1260 ft^3. If the length of the room is 14 ft and the height is 7.5 ft, find the width of the room.

3. Find the volume of a pipe of internal radius 3 cm and length 2.4 m. ($\pi = 3.14$)

4. A block of lead measuring 2 m by 3 m by 1.5 m is melted down to make components as shown. If 3% of the lead is wasted, find the number of components cast.

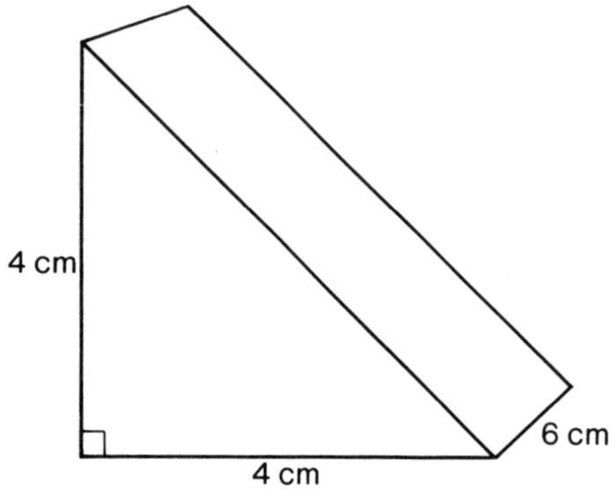

5. A feeding trough for cattle has a triangular cross section with an area of 1800 cm^2. The trough is 2.4 m long. Find the volume of the trough in cubic metres.

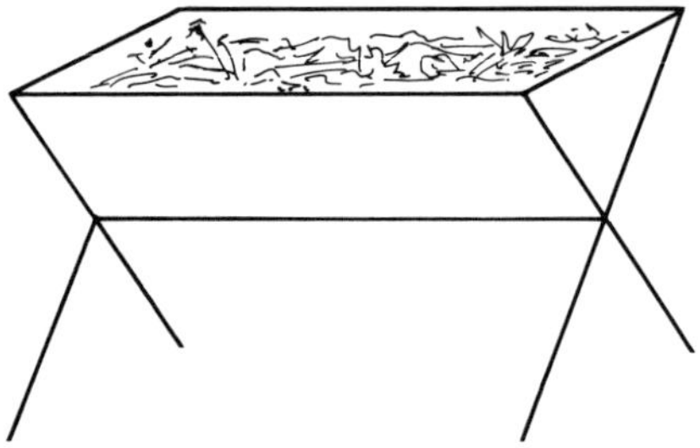

D Find the surface area and volume of these boxes:

1.

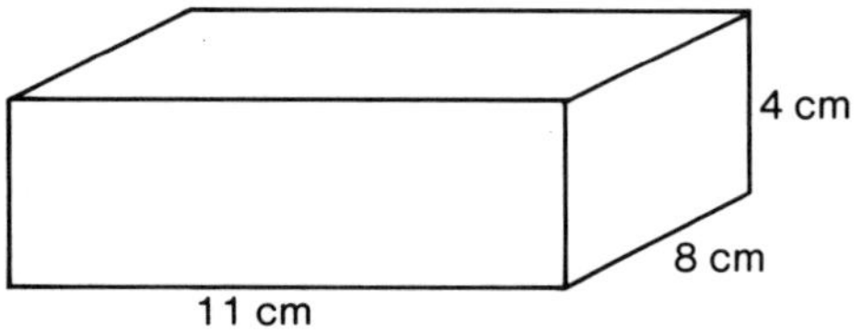

2.

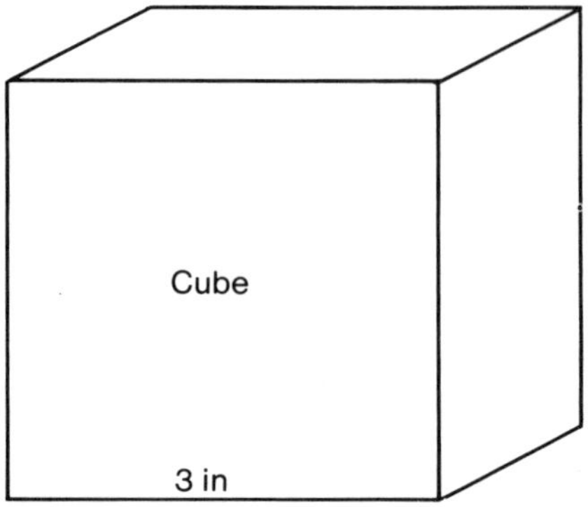

3.

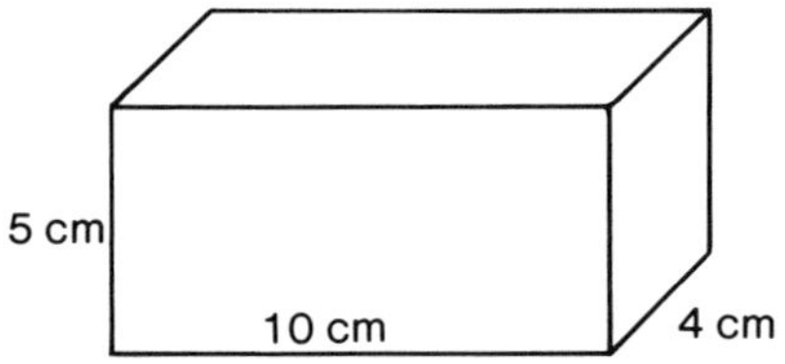

4.

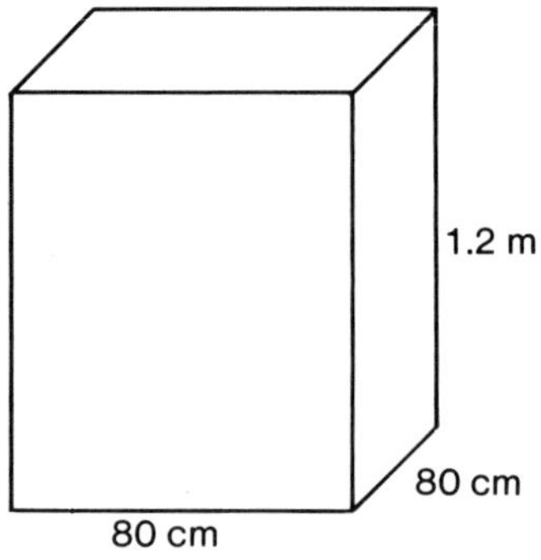

5. A room measures 4.1 m long by 2.8 m wide by 2.7 m high. The walls and ceiling are to be painted with emulsion paint. 1 litre of paint costs £5.95 and covers 15 m^2. Two coats of paint are needed. Ignore the window and door.

a) Find the area to be painted.

b) Find the cost of the paint.

6. A factory produces components as shown. They are painted on all surfaces.

   a) Find the area to be painted on each component.

   b) A tin of paint covers 20 m^2. How many components can be painted with one tin of paint?

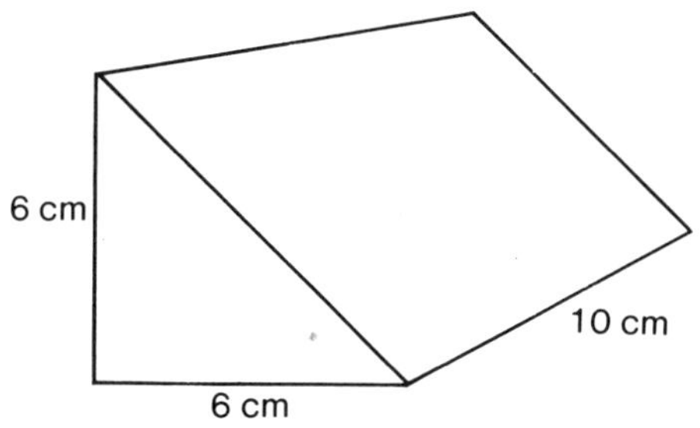

7. A staircase consists of steps each 18 cm high and 24 cm wide. They lead to a landing 2 m long. There are 14 steps.

   a) Find the length of carpet needed to go from the bottom of the stairs to the end of the landing.

   b) If the carpet is 70 cm wide and an extra 50 cm length is bought to allow for wear, find the area of carpet purchased.

   c) The carpet costs £9.85 per square metre. Find the total cost.

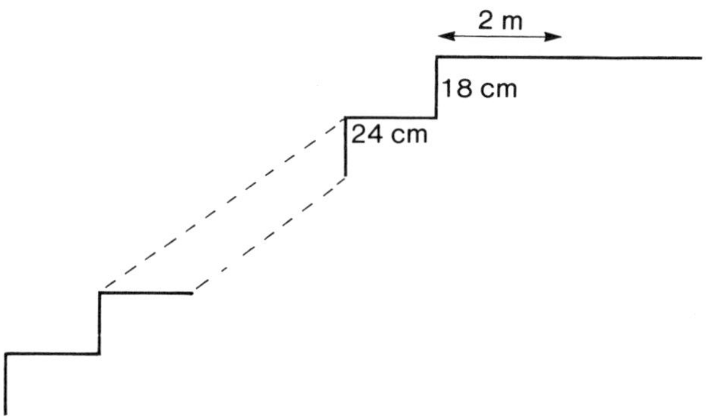

8. A baker is making Christmas cakes. He uses 8 in square tins, which are 3 in deep. He greases the sides and bottom of the tins with fat.

   a) Find the area to be greased in one tin.

   b) If he applies the fat 0.01 in thick, find the volume of fat used on 25 tins.

E

1. Find the capacity of a rectangular tank measuring 40 cm by 30 cm by 20 cm. (Answer in litres.)

2. Find the capacity of a cylinder of radius 8 cm and 15 cm high. ($\pi = 3.14$.) (Answer in litres.)

3. A tin of soup measures 7 cm in diameter and is 11 cm high. Estimate the capacity of the soup. $\left(\pi = \frac{22}{7}.\right)$ (Answer to the nearest 10 millilitres.)

4. A rectangular water tank measures 4.2 m long by 3.5 m wide and is filled to a depth of 1.8 m. Find how many litres of water are in the tank.

5. A bottle contains 1.5 ℓ of medicine. How many 5 mℓ spoonfuls can be taken from it?

6. Mr. Jones has an oil tank measuring 2.5 m by 90 cm by 1.8 m high. He has 19 cm of oil left when the oil tanker arrives. He can only buy oil in units of 100 ℓ.

   a) What is the most oil he can buy without overflowing the tank?

   b) He decides to buy only 3000 ℓ. Find the depth of the oil after the delivery.

   c) The oil is pumped in at 120 ℓ per minute. How long does it take to deliver the oil? (Answer to the nearest minute.)

7. The petrol tank of my car has a capacity of 11 gal. My car is quarter full when I buy another 20 ℓ. (1 gallon = 4.5 ℓ)

a) Find the capacity of my petrol tank in litres.

b) How much petrol is now in the car? Express this as a percentage of a full tank.

c) My car will travel 30 miles per gallon. How far can I now travel before I run out of petrol?

8. A man wishes to lay a concrete path in his garden:

(a) Find the area of the path in square metres.

(b) If the path is 6 cm thick, find the volume of concrete required.

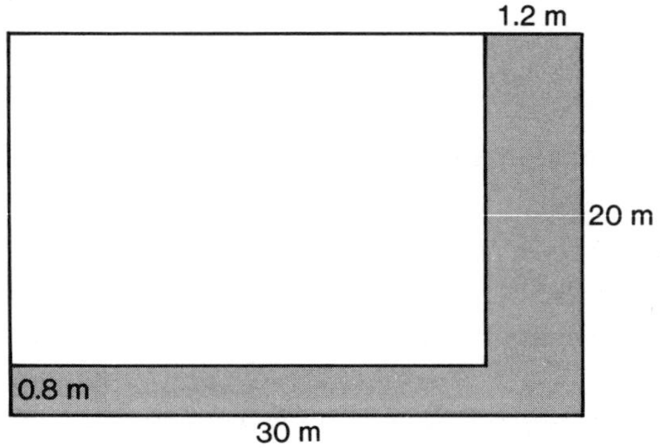

F Exam Questions:

1. A circle is to be cut from a rectangular piece of metal 30 cm long and 21 cm wide.

a) Calculate the area of the largest circle that can be taken from this piece of metal and find the amount wasted. ($\pi = 3\frac{1}{7}$)

b) If 1 cm^2 of the metal weighs 8 g calculate the weight of this circular piece of metal.

c) What percentage of the metal is wasted? (RSA)

2. Mr. Wilson has a rectangular storage tank for central heating oil. It measures 2.4 m long, 80 cm wide and 1.6 m high.

Take 1 litre = 1000 cm^3 and

1 litre = 1.75 pints.

a) How much oil in litres will the tank hold?

b) Calculate the number of gallons held by the tank.

c) How long will it take to fill the empty tank if oil is pumped in at the rate of 120 litres per minute? Give your answer to the nearest minute. (RSA)

3.

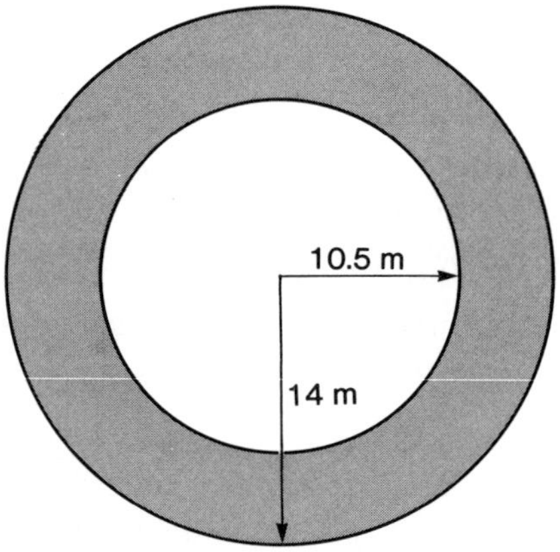

The diagram, which is not drawn to scale, represents a circular garden radius 10.5 m surrounded by a path (shaded in the diagram) with an outer radius of 14 m. Roses costing 64p each are planted in the garden allowing 0.5 m^2 for each rose and the path was covered in tarmac at a cost of £323.40.

a) Find the area of the path. (Take $\pi = 3\frac{1}{7}$)

b) How much does it cost to fill this garden with roses?

c) What was the cost for every square metre of tarmac? (RSA)

4. A family decides to re-decorate their lounge completely. The room involved is 3.9 m long, 3.2 m wide and 3.1 m high.
   a) The walls are painted in emulsion paint which costs £5.85 for a 1 litre tin and each tin is capable of covering 15 m^2. How much will it cost to do this job? The doors and windows are not taken into consideration.
   b) What will be the cost of carpeting the floor, when the carpet used is 70 cm wide and is laid down the *length* of the room? Each metre of carpet used costs £6.75, and is only sold in whole metres.
   c) A rug measuring 1.8 m by 1.6 m is placed on the floor. What percentage of the floor area is not covered by the rug? (Give your answer to the nearest whole number). (RSA)

5.

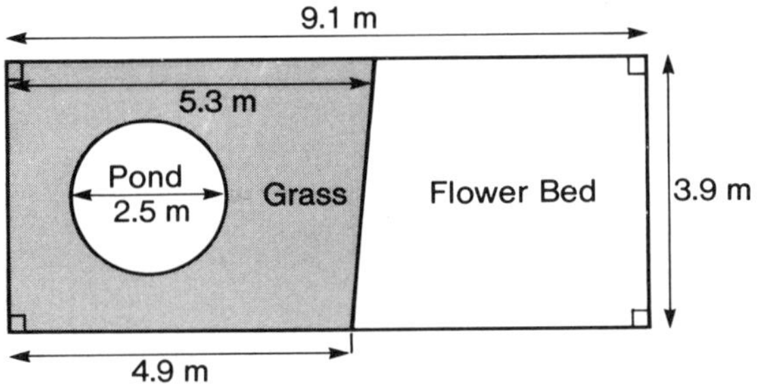

   a) Work out the area of the grass. $\pi = 3.14$ (Give your answer correct to 2 decimal places.)
   b) It is decided to edge the perimeter of the garden and the pond with a low wire fence. If this fencing is sold in 3 m lengths, which cost £5.99 each, how much will the materials cost? ($\pi = 3.14$).
   c) If a garage base measuring 3.2m by 3.4m, was laid on the flower bed, what percentage of the original flower bed would remain? (Give your answer to 1 decimal place.) (RSA)

UNIT NINETEEN

A

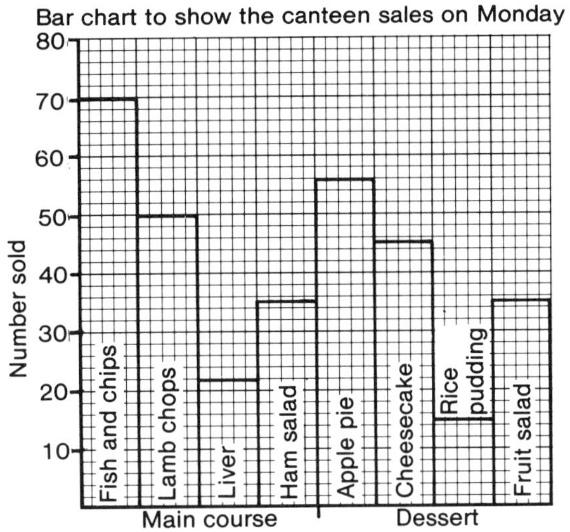

1. How many people had fish and chips?
2. Find the total number of main courses sold.
3. Find the total number of desserts sold.
4. Assuming that everybody who came to the canteen had a main course and that nobody had more than one dessert, find the percentage that did not have a dessert.
5. Find the percentage that had liver.
6. Find the percentage that had fruit salad.
7. On Tuesday the cook prepares chicken pie and meringues instead of liver and rice pudding. The other items are still available. Two less people come to the canteen. 32% choose chicken pie. How many choose chicken pie?
8. Also on Tuesday 11 choose ham salad. The ratio of lamb chops to fish and chips is the same as on Monday. Find the numbers of each main course served on Tuesday.
9. On Tuesday the number of people buying apple pie increases by $12\frac{1}{2}\%$. There is only enough cheesecake for 15 people. 16% do not have a dessert. The ratio of meringue to fruit salad is 2 : 1. Find the numbers of each dessert served on Tuesday.

10. Draw a bar chart to show the canteen sales on Tuesday.

B

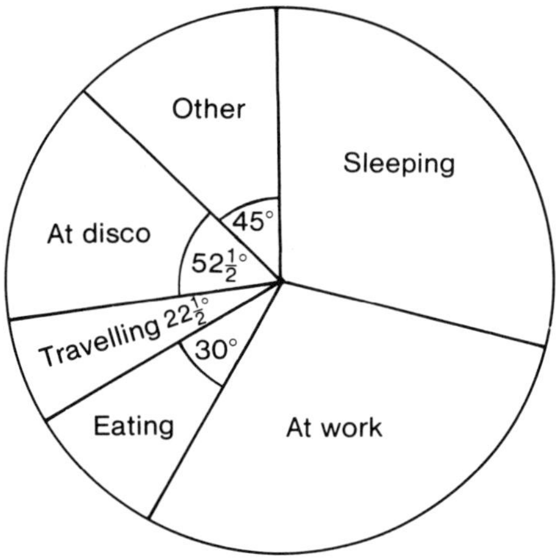

Pie-chart to show how a girl spends her day on Wednesday

1. The girl spends the same length of time at work as she does sleeping. Calculate the number of degrees in the sector labelled sleeping.
2. There are 24 hours in a day. How many degrees on the pie chart represent one hour?
3. How long does the girl spend at work?
4. How long is the girl at home? (eating, sleeping and other)
5. If she arrives at the disco at 8.45 p.m., when does she leave?
6. What percentage of her day is spent at work?
7. What percentage of her time awake is spent travelling?
8. She is paid £2.40 per hour and works 5 days a week. She pays 12% of her earnings in tax and other deductions. How much money does she actually receive after deductions each week?
9. The next day she spends the same amount of time sleeping, at work, eating and travelling as before. During the evening she watches a video for $2\frac{1}{2}$ hours. Draw a pie-chart to show how the girl spends Thursday.
10. On Saturday she spends $8\frac{1}{2}$ hours sleeping, $2\frac{1}{2}$ hours eating, 4 hours shopping, $3\frac{1}{2}$ hours cleaning her flat and 4 hours at a party. Draw a pie-chart to show how she spends Saturday. Label all the angles.

C

A boy collects stamps. He draws a pictogram to show the number of stamps in his collection.

My stamp collection 1984

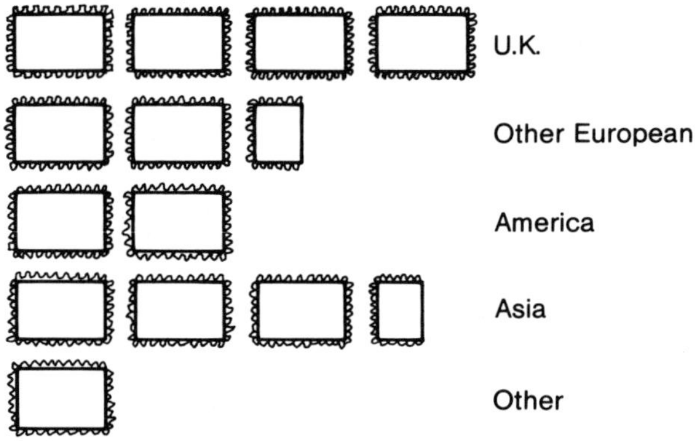

1. He forgets to put a scale on his graph. If he has 1300 stamps altogether, draw his scale.
2. How many United Kingdom stamps does he have?
3. Express the number of stamps from each area in a ratio using whole numbers.
4. In 1985 he has 2080 stamps. The ratio in each section remains the same. How many American stamps does he now have?
5. Draw a pictogram for 1985. Don't forget the scale.

D A group of girls start a parcel delivery service. For short distances they use a bicycle and take parcels up to 5 kg. The charges are shown on line A on the graph. Parcels over 5 kg and all parcels on journeys from 5 to 10 miles are charged on line B. Distances from 10 to 15 miles are charged on line C. All charges are made to the nearest 10 p (exactly 5 p goes up to 10 p).

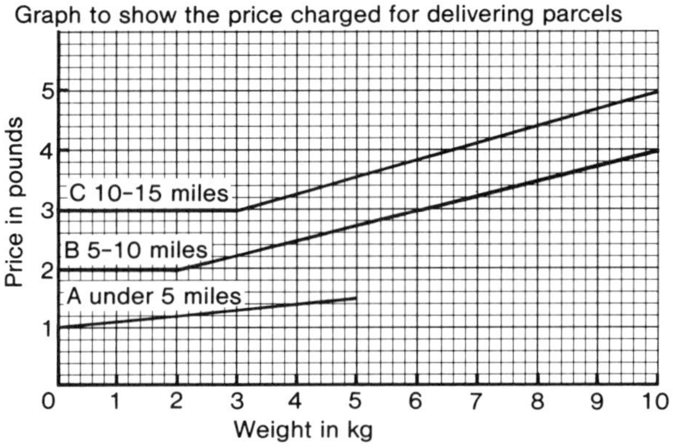

1. How much do they charge to take a parcel of 2 kg a distance of 3 miles?

2. How much do they charge to take a parcel of 8 kg a distance of 12 miles?

3. How much do they charge to take a parcel of 5.5 kg a distance of 4 miles?

4. What is the minimum charge for a distance of 11 miles?

5. A parcel is taken 8 miles. The cost is £2.60. Estimate the weight of the parcel.

6. A parcel is taken 3 miles. The cost is £3.20. Estimate the weight of the parcel.

7. A man wishes to send two parcels weighing 4 kg each to a friend living 8 miles away. Find his percentage saving if he wraps the two parcels together and makes one 8 kg parcel.

8. On a Saturday the girls deliver the following parcels:
   4.5 kg — 2 miles
   2.4 kg — 4.5 miles
   9.0 kg — 6 miles
   1.2 kg — 11 miles
   8.0 kg — 9 miles
   Find the cost of each parcel.
9. Find the average cost of the five parcels delivered on Saturday.
10. Find the total distance the parcels were delivered and the average cost per mile. (Answer to nearest penny.)

E Exam Questions:

1. The following information relates to the running costs of a car for one year.

Petrol	£668
Car Tax	£80
Servicing	£192
Depreciation	£500

   a) Calculate the angles necessary to put this information on to a pie chart.
   b) Construct a pie chart of radius 4 cm.
   c) What is the average weekly cost of running a car? (52 weeks = 1 year) (RSA)

2. The monthly takings of a small business for the first eight months of the year are as follows:-

Jan.	Feb.	March	April
£3500	£4300	£3850	£4600

May	June	July	August
£4450	£4800	£4250	£4050

   a) Draw a bar chart to illustrate the figures using 2cm to represent £500 on the vertical scale.
   b) Find the average monthly takings and show this on the chart by drawing a straight line.
   c) If the average monthly takings for the next four months are £4562, find the annual takings of this business. (RSA)

3. The pie chart below shows the relative costs of manufacturing an article which sells for £120.

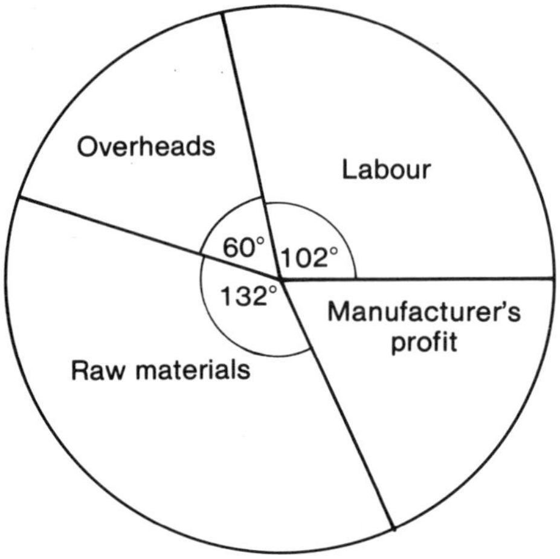

a) Find the cost of making an article.
b) If the labour costs are reduced by 10%, the raw material costs are increased by 5% and the overheads remain the same, find the new cost of making this article.
c) If the selling price remains the same, express the new profit as a % of this selling price giving your answer correct to the nearest whole number. (RSA)

4. A student decides to brew 9 litres of brown ale at home. The ingredients he will need to use are listed below:

BROWN ALE

555g DRIED MALT EXTRACT
125g CRUSHED CRYSTAL MALT
50g CRUSHED BLACK MALT
225g SOFT BROWN SUGAR
225g BREWING SUGAR
20g HOPS

a) Work out the size of each angle which would represent each of the ingredients on a pie chart.
b) Draw a neat, accurate pie chart using your answers from part (a). The radius of the pie chart should be 1¾ inches.
c) A brewery uses the same recipe but on a far larger scale. It has six 27000 litre vessels. How many kilogrammes of the ingredients will be used, in total, if each vessel is in use for 336 days of each year? The brewing process for each vessel takes 21 days. (RSA)

5.

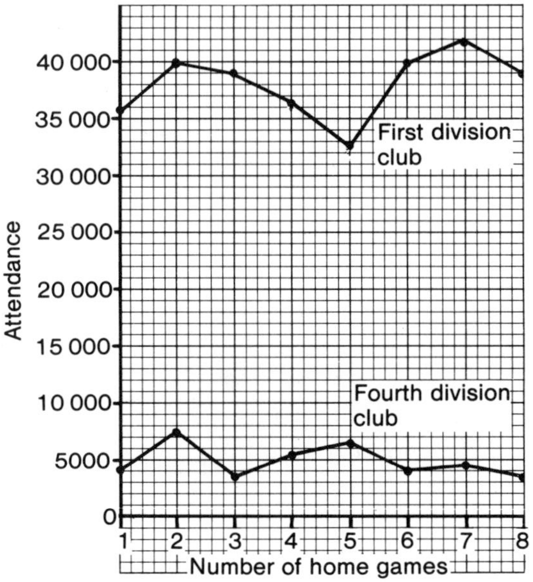

The graph shows the attendances over a period of EIGHT home games of a FIRST DIVISION FOOTBALL CLUB and a FOURTH DIVISION FOOTBALL CLUB.

a) Find the average weekly attendance of the Fourth Division Club.

b) What is the largest variation of attendance between the two clubs and for which week does it occur?

c) Express the attendance in week 2 of the Fourth Division Club as a percentage of the First Division Club in the same week giving your answer correct to one decimal place. (RSA)

UNIT TWENTY

Here are some more questions from old exam papers for you to try.

1. A man and his wife take a 14 day holiday in Majorca. The cost for each adult is £315 which includes half board (breakfast and dinner). There is a supplement (an additional payment) for full board (breakfast, lunch and dinner) of £2 a day each and a room with a sea view costs an additional £1.50 a day each.

   a) What is the total cost of his holiday to the man and wife if they decide to have full board and a room with a sea view?

   b) If they change £215 to pesetas at an exchange rate of 192.50 pesetas to the £, how many pesetas will they receive?

   c) At the end of the holiday they change 1980 pesetas back to £ at an exchange rate of 195 pesetas to the £. Calculate how much they will receive if the bank charges 50p commission for the transaction. (Give your answer to the nearest penny.) (RSA)

2. A man travelled by car from Manchester to Newcastle, a distance of 130 miles. The journey took 3 hours 35 minutes and included a stop of 15 minutes for a coffee.
   The car left Manchester at 11.30 and travelled at an average speed of 48 mph until the stop for coffee was made at 12.45.

   a) What was his average speed for the second part of the journey?

   b) The average petrol consumption figures for his car are as follows:

      30 to 40 mph   33 mpg
      40 to 50 mph   27.8 mpg

      Work out how much petrol he used on the trip. (Give your answer correct to 2 decimal places.)

   c) The petrol he used cost £1.85 per gallon. As the man owns his own business he will be able to claim back the V.A.T. which is included in the price of the petrol. (V.A.T. Rate 15p in the £1.) How much V.A.T. will he claim back for this trip? (Give your answer to the nearest penny.) (RSA)

3. The rateable value of a house in Dimshire is £350. The annual rate in the £ is 125.9p. Out of the total rates received from the householder the local parish received £13.30.

   a) Find the average monthly rate giving your answer correct to the nearest penny.

   b) What is the local rate in the £?

   c) Find what percentage the local annual rate is of the total annual rate giving your answer correct to 2 decimal places. (RSA)

4. A man is paid £3.86 per hour for a 40 hour week. Any overtime work during the week is paid at time and a half the normal rate. He is paid double time for any Saturday work. During a particular week he works the following hours:

Monday	8½
Tuesday	10
Wednesday	9½
Thursday	9
Friday	8
Saturday	4

   a) How much would he earn for this week and how much would his *net* wage be if his total deductions amount to 31% of his gross wage? (Answers to the nearest penny).

b) How many hours and minutes of overtime would he need to work in order to earn the same money he would earn during 5 hours of normal work? (The overtime involved is not at the Saturday rate.)

c) If his normal hourly rate is increased to £4.20, what percentage pay rise will he receive? (Give your answer correct to 1 decimal place.) (RSA)

5. A kitchen floor measures 12′ long by 8′ 10½″ wide. It is to be covered with 6″ square heavy duty ceramic tiles. A ¼″ gap is left between each tile and around the edge of the floor.

a) How many tiles will be required to complete the job?

b) If the tiles are only sold in boxes of 20, costing £5.84 per box. How much will it cost to tile the floor?

c) The gaps between the tiles and around the edges are filled with grout. What area of the floor in square inches will be exposed grout? (RSA)

6. Mr. Jones finds that he can buy a television set for £355 which is guaranteed for the first two years against any faults. For the next five years he can pay an annual fee of £37.35 to cover all maintenance. He could rent the same set at a cost of £7.24 a month and this would include all maintenance.

a) Find how much Mr. Jones would pay over the seven years if he decided to purchase the television set and pay the 5 years maintenance fee.

b) How much would it cost Mr. Jones to rent the same set over the 7 year period?

c) Mr. Jones buys a set. If he sells this set at the end of the seven year period for £50, how much will he have saved by buying the television rather than renting it? (RSA)

7. In 1980 the cost of producing a machine was divided amongst labour, materials and overheads in the ratio of 5 : 3 : 2. In 1982 the cost of materials had risen 3 times as much and the overheads twice as much but the labour costs were reduced by 25% of the 1980 figures. The cost of the machine in 1980 was £2560.

a) Find the cost of the labour, materials and overheads in 1980.

b) What was the cost of producing this machine in 1982?

c) In 1982 the manufacturer made a profit of 15% on his cost price, find how much profit he would make if he sold 26 machines. (RSA)

8. A lady received a statement from her Building Society showing that her capital had gained interest of £47.50, during the year.

a) If the rate of interest was 9½% per annum how much capital did she have in the bank at the beginning of that year?

b) The lady inherits £3500 and decides to invest this sum in a special account available at her Building Society. This account pays interest at 11% per annum, but her money must stay invested in the account for three years. How much will she have in her account at the end of the three year period? (The interest is calculated as compound interest and taken to the nearest penny).

c) Express the increase in her capital at the end of the three year period as a percentage of the original capital. (Give your answer correct to 3 decimal places.) (RSA)

9. The following items were purchased from a supermarket in 1982 and then again in 1983. The prices are given:

	1982	*1983*
SUGAR (1 kg)	46 p	52 p
FLOUR (1.5 kg)	37 p	39 p
CAN SOUP (435 g)	21 p	24 p
BUTTER (250 g)	40 p	44 p
TEA BAGS (500 g)	£1.52	£1.59
TOMATO KETCHUP (425 g)	40 p	42 p

a) Express the total cost of the items in 1982 as a fraction of the total cost in 1983. Give your answer in its lowest terms.

b) What is the total weight, in kilogrammes, of the items listed?

c) Which of the items in 1983 has risen by exactly 10% of its 1982 price?

d) In 1984 the price of sugar is expected to be 59 p per kilogramme. By how much is this percentage rise larger than for the previous year? (Work to 2 decimal places.) (RSA)

Answers to Extra Examples

UNIT ONE

A
1. 27 2. 75
3. 101 4. 201
5. 250 6. 726
7. 935 8. 503
9. 981 10. 292

B
1. 23 2. 44
3. 132 4. 19
5. 159 6. 787
7. 164 8. 358
9. 138 10. 2521

C
1. 72 2. 18
3. 21 4. 56
5. 45 6. 81
7. 48 8. 20
9. 28 10. 18

D
1. 6 2. 4
3. 9 4. 9
5. 4 6. 9
7. 9 8. 6
9. 8 10. 8

E
1. 144 2. 243
3. 100 4. 80
5. 684 6. 1505
7. 760 8. 333
9. 920 10. 1782

F
1. 90 2. 40
3. 1000 4. 920
5. 8400 6. 89 000
7. 1400 8. 62 000
9. 4160 10. 518 000

G
1. 300 2. 760
3. 17 200 4. 204
5. 2378 6. 3196
7. 5402 8. 16 932
9. 46 080 10. 82 194

H
1. 220 2. 490
3. 194 4. 256
5. 981 6. 952
7. 521 8. 897
9. 654 10. 658

I
1. 2 2. 10
3. 55 4. 200
5. 34 6. 80
7. 2760 8. 51
9. 8760 10. 310

J
1. 15 2. 13
3. 64 4. 56
5. 75 6. 96
7. 114 8. 245
9. 254 10. 999

K
1. 14, 15 2. 21, 20
3. 32, 34 4. 18, 20
5. 13, 15 6. 22, 27
7. 8, 3 8. 48, 56
9. 21, 14 10. 31, 63

L

1.	808	2.	90
3.	721	4.	1242
5.	9	6.	44
7.	8	8.	718 000
9.	10 764	10.	979
11.	211	12.	35, 41
13.	20	14.	1260
15.	189	16.	1292
17.	8	18.	1208
19.	896	20.	172

UNIT TWO

A

1.	12.8	2.	12.3
3.	27.07	4.	5.9
5.	18.18	6.	8.89
7.	44.8	8.	13.8
9.	206.72	10.	193.28

B

1.	58	2.	8360
3.	2100	4.	890
5.	6710	6.	2
7.	0.361	8.	35 920
9.	17 210	10.	10 190

C

1.	49	2.	7.4
3.	0.12	4.	20.52
5.	72.21	6.	3.472
7.	27.03	8.	0.259 14
9.	563.52	10.	1943

D

1.	0.379	2.	0.1638
3.	0.08	4.	0.1274
5.	0.1572	6.	2.17
7.	1.59	8.	0.027 92
9.	0.2789	10.	0.28

E

1.	9	2.	5.6
3.	2.54	4.	96.8
5.	7.01	6.	2.561(7)
7.	52.000	8.	24.705(8)
9.	7.500	10.	30.439(0)

F

1.	2.2	2.	3.5
3.	1.666(6)	4.	2.666(6)
5.	3	6.	2.5
7.	5.25	8.	2.375
9.	0.375	10.	3.285(7)

G

1.	1.94	2.	1.44
3.	1570	4.	0.28
5.	35	6.	2.571(4)
7.	5.91	8.	24.05
9.	0.1	10.	106.222(2)
11.	18.25	12.	20.4
13.	860	14.	0.1765
15.	84	16.	7.95
17.	1.62	18.	67.86
19.	26.69	20.	0.2839

UNIT THREE

A

1. $\frac{1}{4}$

2. $\frac{3}{5}$

3. $\frac{5}{6}$

4. $\frac{4}{6}$ or $\frac{2}{3}$

5.

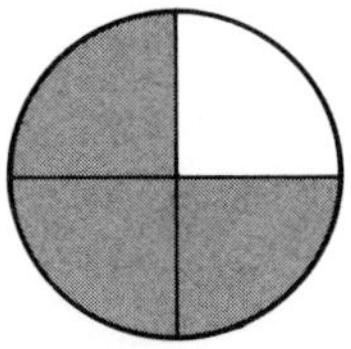

6.

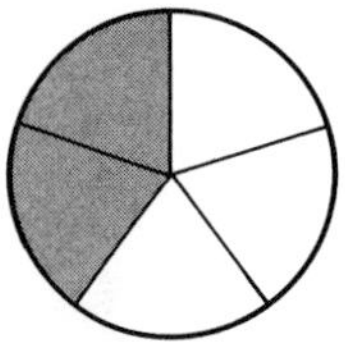

7.

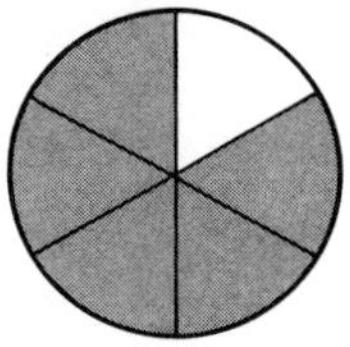

8.

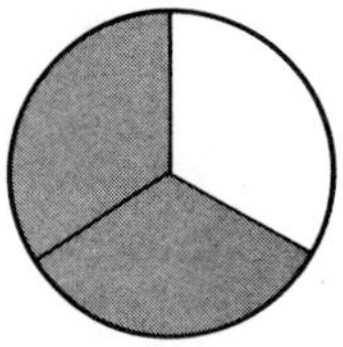

9.

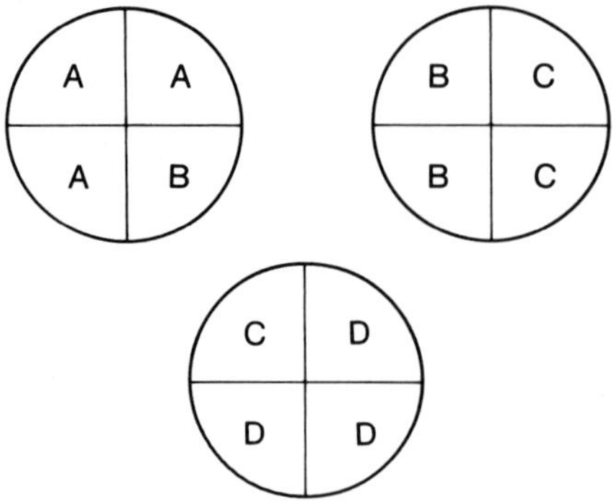

They get $\frac{3}{4}$ each

10.

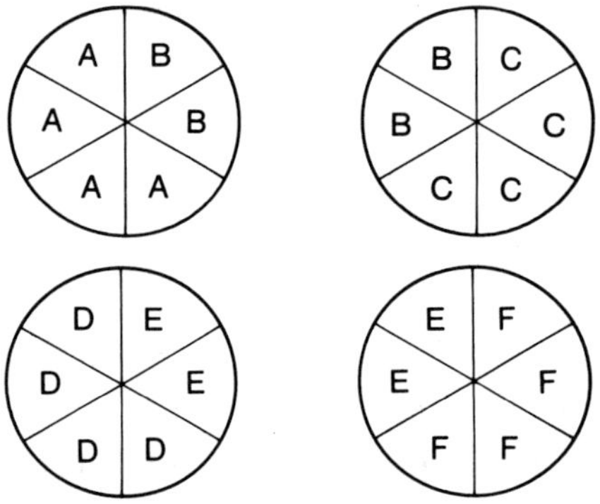

They get $\frac{4}{6}$ each $\left(\text{or } \frac{2}{3}\right)$

B

1.	$\frac{1}{2}$	**2.**	$\frac{2}{3}$	**3.**	$\frac{3}{5}$
4.	$\frac{2}{3}$	**5.**	$\frac{1}{2}$	**6.**	$\frac{1}{3}$
7.	$\frac{3}{7}$	**8.**	$\frac{1}{3}$	**9.**	$\frac{3}{4}$
10.	$\frac{9}{17}$	**11.**	$\frac{4}{6}$	**12.**	$\frac{9}{12}$
13.	$\frac{3}{9}$	**14.**	$\frac{33}{36}$	**15.**	$\frac{8}{20}$
16.	$\frac{5}{20}$	**17.**	$\frac{6}{9}$	**18.**	$\frac{3}{21}$
19.	$\frac{18}{30}$	**20.**	$\frac{60}{80}$	**21.**	$\frac{11}{2}$
22.	$\frac{10}{3}$	**23.**	$\frac{9}{2}$	**24.**	$\frac{5}{3}$
25.	$\frac{25}{4}$	**26.**	$\frac{13}{3}$	**27.**	$\frac{37}{4}$
28.	$\frac{51}{5}$	**29.**	$\frac{37}{3}$	**30.**	$\frac{127}{6}$
31.	$2\frac{3}{4}$	**32.**	5	**33.**	$2\frac{4}{7}$
34.	$3\frac{1}{5}$	**35.**	$5\frac{1}{4}$	**36.**	$8\frac{1}{3}$
37.	$8\frac{1}{4}$	**38.**	$7\frac{3}{5}$	**39.**	$3\frac{3}{8}$
40.	$9\frac{3}{10}$				

C

1.	4	**2.**	$4\frac{3}{4}$	**3.**	$5\frac{8}{15}$
4.	$11\frac{1}{4}$	**5.**	$6\frac{1}{3}$	**6.**	$12\frac{5}{6}$
7.	$12\frac{8}{9}$	**8.**	$8\frac{1}{15}$	**9.**	$169\frac{7}{9}$
10.	$47\frac{50}{63}$				

D

1.	$2\frac{1}{4}$	**2.**	$5\frac{1}{6}$	**3.**	$4\frac{10}{21}$
4.	$4\frac{7}{12}$	**5.**	$1\frac{1}{12}$	**6.**	$4\frac{7}{24}$
7.	$\frac{9}{16}$	**8.**	$4\frac{1}{6}$	**9.**	$8\frac{61}{63}$
10.	$91\frac{1}{36}$				

E

1.	$2\frac{1}{4}$	**2.**	15	**3.**	$10\frac{2}{3}$
4.	12	**5.**	$2\frac{1}{2}$	**6.**	$17\frac{2}{5}$
7.	6	**8.**	$2\frac{1}{4}$	**9.**	$\frac{2}{9}$
10.	$7\frac{1}{2}$				

F

1.	1	**2.**	$\frac{1}{2}$	**3.**	$\frac{5}{6}$
4.	$\frac{8}{9}$	**5.**	$\frac{2}{3}$	**6.**	$1\frac{1}{2}$
7.	$7\frac{1}{8}$	**8.**	$\frac{92}{99}$	**9.**	12
10.	$4\frac{1}{2}$				

G

1.	$\frac{2}{5}$	**2.**	$\frac{3}{4}$	**3.**	$2\frac{2}{5}$
4.	$6\frac{17}{20}$	**5.**	$9\frac{1}{4}$	**6.**	$4\frac{1}{8}$
7.	$6\frac{3}{8}$	**8.**	$21\frac{9}{20}$	**9.**	$14\frac{13}{40}$
10.	$12\frac{19}{20}$				

H

1. 0.75 2. 0.375
3. 0.4 4. 0.444(4)
5. 0.666(6) 6. 9.6
7. 8.428(5) 8. 12.8
9. 0.785(7) 10. 2.652(1)

I

1. 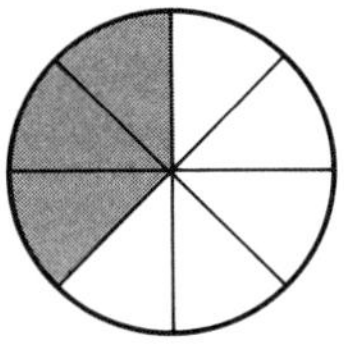 2. $6\frac{1}{12}$

3. $3\frac{2}{5}$ 4. $\frac{8}{12}$
5. $1\frac{7}{20}$ 6. $4\frac{5}{6}$
7. $2\frac{5}{8}$ 8. 7.375
9. $38\frac{4}{15}$ 10. $1\frac{3}{4}$
11. $1\frac{5}{8}$ 12. 8
13. $3\frac{7}{8}$ 14. $3\frac{1}{3}$
15. $8\frac{1}{3}$ 16. $3\frac{3}{5}$
17. $1\frac{5}{12}$ 18. $1\frac{2}{3}$
19. $6\frac{5}{12}$ 20. $\frac{1}{2}$

UNIT FOUR

A

1. 512 p 2. 627 p
3. 1200 p 4. 904 p
5. 830 p 6. £1.25
7. £6.38 8. £13.12
9. £0.29 10. £0.05

B

1. £1.23 2. £27.93
3. £4.26 4. 4.61 marks
5. 20.82 francs 6. $7.87
7. 18 p 8. 130.50 francs
9. £1.29 10. $44.25

C

1. £18.90 2. £21.15
3. £8.54 4. £8.26
5. $31.92 6. 13.20 francs
7. £13.50 8. £31.05
9. £10.75 10. $44.25

D

1. 27 p 2. £1.06
3. £1.70 4. £1.60
5. $0.65 or 65 cents 6. 2.33 francs
7. 2.75 marks 8. £366.50
9. £1.25 10. £5.87

E

1. 108 francs 2. 175 marks
3. 44 000 lira 4. $220
5. 8000 pesetas 6. £2
7. £200 8. £3
9. £4.75 10. £800

F

1. 1296 francs 2. £20
3. 88 000 lira 4. £12.50
5. $5.50 6. £455
7. £1.50 8. 900 pesetas
9. £53 10. £125

G

1. £15.35 2. £4.23
3. £2.30 4. $99.20
5. 2484 francs 6. £6.41
7. £6.48 8. 23
9. £5.50 10. £104
11. £78.48 12. £2.35
13. 1687.5 pesetas 14. £3.68
15. £4900 16. $397.50
17. £2430 18. £2.34
19. 5 20. 464.60 marks

UNIT FIVE

A

1.	100	**2.**	1000
3.	100	**4.**	km
5.	litre	**6.**	kg
7.	1000	**8.**	mm
9.	cℓ	**10.**	cm

B

1.	3	**2.**	foot
3.	oz	**4.**	8
5.	36	**6.**	mile
7.	16	**8.**	gallon
9.	12	**10.**	1760

C

1. 28 mm = 2 cm 8 mm
2. 12 500 g = 12 kg 500 g
3. 2.412 km = 2 km 412 m
4. 8.62 ℓ = 8 ℓ 620 mℓ
5. 23.8 m = 23 m 80 cm
6. 2 ft 1 in
7. 5 yd 2 ft
8. 4 lb 9 oz
9. 5 gal 4 pt = 44 pt
10. 1 lb 12 oz = 28 oz

D

1.	7 m 7 cm	**2.**	5.6 m
3.	9.75 m	**4.**	6.5 kg
5.	3.83 kg	**6.**	1 ℓ
7.	3500 g	**8.**	4.31 kg
9.	180 cm	**10.**	1400 mℓ

E

1.	19.2 ℓ	**2.**	23.2 kg
3.	50 cℓ	**4.**	400 m
5.	3.5 kg	**6.**	21.6 cm
7.	18 m	**8.**	1875 m
9.	54 kg	**10.**	125 mℓ

F

1.	7 ft 8 in	**2.**	5 pt
3.	7 lb 14 oz	**4.**	352 yd
5.	1 gal 7 pt	**6.**	1 lb 10 oz
7.	$14\frac{1}{2}$ oz	**8.**	20
9.	14 stone 2 lb	**10.**	117 gal

G

1.	20.32 cm	**2.**	8 km
3.	40.5 ℓ	**4.**	3 ft 2 in
5.	20 kg	**6.**	20 gal
7.	51.2 km	**8.**	71.12 cm
9.	110 lb	**10.**	62.5 mile

H

1.	11 ft	**2.**	4.268 m
3.	602 cm	**4.**	7 in
5.	76.5 ℓ	**6.**	16 ft 8 in
7.	2500 g	**8.**	6.5 m
9.	2.52 ℓ	**10.**	680 cℓ
11.	42 m 60 cm	**12.**	49.5 kg
13.	12 ℓ	**14.**	11 m 34 cm
15.	10 500 g	**16.**	28 ℓ
17.	52 m 50 cm	**18.**	20.02 m
19.	33.75 kg	**20.**	104 gal

UNIT SIX

A

1.	£5.28	**2.**	87 p
3.	£10.75	**4.**	£12.10
5.	93 p	**6.**	£0.04
7.	10 p	**8.**	£7.17
9.	£24.72	**10.**	62 p

B

1.	15.91	**2.**	20.7
3.	0.005	**4.**	20.0
5.	0.011	**6.**	27.94
7.	10.0	**8.**	0.0
9.	11.6	**10.**	72.85

C

1. 19.2
2. 24
3. 105
4. 92.9
5. 21.21
6. 0.0147
7. 0.028
8. 201
9. 19 800
10. 2200 000

D

1. 1.875
2. 1.8
3. $1\frac{4}{5}$, 1.85, $1\frac{7}{8}$
4. 15
5. 16
6. $\frac{17}{20}, \frac{4}{5}, \frac{3}{4}$
7. 9.51
8. $3^2, \frac{19}{2}, \sqrt{100}$
9. $2\frac{1}{5}$
10. $3\frac{1}{7}$, 3.14, $3\frac{2}{15}$

E

1. 81
2. $2\frac{1}{4}$
3. 4.41
4. 900
5. 40 000
6. 5
7. 8
8. 100
9. 0.5
10. $\frac{1}{3}$

F

1. 4
2. 12
3. 1
4. 12
5. 10
6. 11
7. 8
8. 1
9. 5
10. 7

G

1. 4
2. 12
3. 56
4. 8
5. 60
6. 30
7. 120
8. 44
9. 30
10. 28

H

1. £9.49
2. 8.72
3. 8.720
4. 4.99
5. 14
6. 3
7. £4.22
8. 31.00
9. 12
10. 6.76
11. $\frac{11}{3}$, 3.6, $3\frac{5}{9}$
12. 42
13. $\frac{8}{3}$, 2.6, $2\frac{2}{9}$
14. 3500 miles
15. $2\frac{3}{4}$, 2.7, $\frac{13}{5}$
16. $2\frac{2}{3}$
17. 60
18. $12\frac{1}{4}$
19. 19 000
20. 1.1

UNIT SEVEN

A

1. $62\frac{1}{2}\%$
2. 40%
3. $87\frac{1}{2}\%$
4. $66\frac{2}{3}\%$
5. 30%
6. 250%
7. 75%
8. 90%
9. 35%
10. 46%

B

1. $\frac{4}{5}$
2. $\frac{1}{2}$
3. $\frac{9}{20}$
4. $\frac{4}{25}$
5. $1\frac{1}{4}$
6. $\frac{1}{8}$
7. $\frac{11}{40}$
8. $\frac{1}{12}$
9. $1\frac{1}{3}$
10. $3\frac{1}{2}$

C

1. 85%
2. 68%
3. 25%
4. $6\frac{1}{4}\%$
5. 20%
6. 1%
7. 20%
8. $13\frac{1}{3}\%$
9. 30%
10. $33\frac{1}{3}\%$

D

1. £10 2. £32
3. 1 kg 4. 1 ℓ 500 mℓ
5. 15 m 6. £4.80
7. £9 8. £17.25
9. 27 10. 770 ℓ

E

1. £40 2. £288
3. £2400 4. $750
5. 450 marks 6. £4
7. £100 8. £500
9. £5900 10. 60 p

F

1. £120 2. £10
3. £414 4. £109.82
5. £8280 6. £60, 15%
7. £1, 5% 8. 10%
9. 1 p 10. 15%

G

1. $137\frac{1}{2}$% 2. $\frac{17}{20}$
3. 8% 4. £45
5. 36 kg 6. £48
7. £3.60 8. $3\frac{1}{3}$%
9. £59.70 10. 22%
11. £1.05 12. $87\frac{1}{2}$%
13. 4% 14. £202.50
15. £1.13 16. £615.30
17. 4% 18. $26\frac{2}{3}$%
19. 67.5 p 20. 20%

UNIT EIGHT

A

1. 23 2. 153.75
3. £1.59 4. 812 g
5. 18.9° 6. 2 m 54 sec
7. 14.2 p 8. 24.3 cm
9. 0.857 10. £45 400

B

1. 185 2. 676
3. 19.45 4. 9.075
5. $\frac{1}{2}$ 6. $6\frac{17}{24}$
7. 740 8. £12.25
9. $\frac{13}{20}$ 10. £150
11. £4.20 12. £137.12
13. 15 14. 25%
15. £126 16. £150
17. £54 18. 6600
19. 19.2 20. 322 miles

UNIT NINE

A

1. £13.50 2. £4.50
3. £1.32 4. £1.14
5. 15 6. $5\frac{1}{3}$ weeks
7. £7.50 8. £35.70
9. 9 10. £535

B

1. 3 : 4 2. 3 : 7
3. 2 : 3 4. 5 : 4
5. 9 : 19 6. 2 : 3
7. 3 : 8 8. 9 : 14
9. 8 : 11 10. 2 : 1

C

1. £8 : £12 2. £75 : £25
3. £16 : £20 4. £24 : £48
5. £63 : £45 6. 10 : 20 : 20
7. 70 : 80 8. 250 g
9. £12 500 10. 350 kg

D

1. 60 2. 200 g
3. £1000 : £2000 4. 6 lb : 10 lb
5. 24 and 40 6. 96
7. 6 : 5 8. $4\frac{1}{2} : 9 : 13\frac{1}{2}$ gal
9. 60 p : 66 p : 78 p 10. 150 g

E

1. 21 p	2. £2.08
3. 19 p	4. £3.74
5. £5.60	6. 61 p
7. 24 p	8. £1.97
9. 33 p	10. 47 p

F

1. £10.80	2. 10 : 17
3. £60 : £30	4. £180
5. £1.02	6. £6.78
7. £500 : £625 : £375	8. 2 : 1
9. 2 : 3	10. 200
11. £135 : £90	12. 135 : 108
13. £23	14. 117 : 195
15. £355 : £284	16. £40
17. 80	18. £27.50
19. 360	20. £36 : £48

UNIT TEN

A

1. 9 cm^2	2. 8 cm^2
3. 9 cm^2	4. $12\frac{1}{2}$ cm^2

B

1. 100	2. 10 000
3. 1000 000	4. 1000 000
5. 144	6. 9
7. 0.16	8. 2
9. 10 280	10. 0.2758

C

1. 16 cm^2	2. 6 in^2
3. 36 cm^2	4. $11\frac{1}{4}$ cm^2
5. 13.6 m^2	6. 1036 in^2
7. $1\frac{1}{4}$ m^2	8. $12\frac{1}{4}$ in^2
9. 70.56 mm^2	10. 10 800 cm^2

D

1. 4 cm^2	2. 1 in^2
3. 5 cm^2	4. $1\frac{7}{8}$ in^2
5. 10 m^2	6. 21.06 m^2
7. 6 ft^2	8. 16.5 m^2
9. 3600 cm^2	10. 150 cm^2

E

1. 154 m^2	2. 616 cm^2
3. 2464 mm^2	4. $38\frac{1}{2}$ cm^2
5. 15 400 m^2	6. 12.56 m^2
7. 78.5 cm^2	8. 7.065 cm^2
9. 50.24 in^2	10. 28.26 ft^2

F

1. 3 cm	2. 5 cm
3. 27.5 cm	4. 6 cm
5. 6 m	6. 7 m
7. 12.5 m	8. 18 cm
9. 64	10. 8

G

1. 20 000 cm^2	2. 720 in^2
3. $29\frac{3}{4}$ cm^2	4. $17\frac{1}{9}$ cm^2
5. $6\frac{1}{4}$ m^2	6. 12.5 m
7. 6.1544 m^2	8. 0.52 km^2
9. $3\frac{3}{4}$ m	10. 7546 cm^2
11. 5544 cm^2	12. 6.4 m
13. 616 cm^2	14. 6 cm
15. £60	16. £720
17. 113.04 m^2	18. 154 cm^2
19. 12 cm	20. $38\frac{1}{2}$ cm^2

UNIT ELEVEN

A

1. 14 cm	2. 10 cm
3. 14 cm	4. 17 cm
5. 16 cm	6. 31.4 m
7. 1 ft	8. 6.2 m
9. 39 ft 0 in	10. 90 cm

B

1.	16 m^2	2.	50 cm^2
3.	20 in	4.	26 m
5.	13.6 m	6.	225 m^2; 200 m^2
7.	20 ft; 50%	8.	£648
9.	300%	10.	6 m

C

1.	44 cm	2.	88 cm
3.	176 cm	4.	440 m
5.	1 ft 10 in	6.	12.56 m
7.	37.68 cm	8.	15.7 m
9.	31.4 in	10.	18.84 ft

D

1.	14 cm	2.	$3\frac{1}{2}$ m
3.	2464 m^2	4.	3.14 m^2
5.	44 cm	6.	88 m
7.	12.56 cm	8.	314 m^2
9.	$38\frac{1}{2}$ ft	10.	159 m

E

1.	24 cm^3	2.	60 cm^3
3.	58.8 cm^3	4.	27 in^3
5.	30 m^3	6.	2 cm
7.	2 cm	8.	7 ft 6 in
9.	16 cm	10.	11 250 ℓ

F

1.	28.4 m	2.	625 ft^2
3.	50.24 ft^2	4.	125.6 cm
5.	61 600 m^2	6.	3600 cm^3
7.	60 000 ℓ	8.	2826 cm^2
9.	$13\frac{6}{7}$ cm	10.	21%
11.	33 cm	12.	36 cm
13.	520 cm^3	14.	177.8 yd^2
15.	220 cm	16.	25 yd^2
17.	70 m	18.	176 cm
19.	£94.60	20.	225 m^2

UNIT TWELVE

A

1.	0315	2.	1925
3.	2024	4.	0613
5.	0004	6.	2.15 a.m.
7.	9.28 a.m.	8.	9.12 p.m.
9.	11.19 p.m.	10.	12.55 p.m.

B

1.	2 h 14 m	2.	55 m
3.	1 h 52 m	4.	1 h 21 m
5.	8 h 50 m	6.	9 h 45 m
7.	20 h 10 m	8.	4 h 55 m
9.	24 h 37 m	10.	3 h 25 m

C

1.	0919; 1 h 51 m	2.	15 m
3.	6	4.	0813
5.	0836; 38 m	6.	20 m
7.	1 h 33 m	8.	2 h 44 m
9.	1 h 41 m		
10.	1 h 27 m; Basingstoke		

D

1.	3 h	2.	5 m.p.h.
3.	1 h 30 m	4.	3 m/s
5.	$37\frac{1}{2}$ m.p.h.	6.	40 km
7.	300 miles	8.	40 m.p.h.
9.	15 m	10.	3 h

E

1.	$16\frac{2}{3}$ m/s	2.	1440 km/h
3.	$133\frac{1}{3}$ cm/s	4.	60 m.p.h.
5.	60 m.p.h.	6.	3 miles
7.	20 km/h	8.	600 m.p.h.
9.	1 h 36 m	10.	390 m

F

1. £36
2. £154
3. £3.00, £4.00, £4.50, £6.00
4. £54
5. £114
6. £92
7. £156
8. £79.50
9. £141
10. £162

G

1. 1407
2. 5.59 p.m.
3. 38 m
4. 3 h 19 m
5. 80 m
6. $33\frac{1}{3}$ m.p.h.
7. $3\frac{3}{5}$ m.p.h.
8. £90
9. £1180
10. 121 miles
11. $25\frac{3}{5}$ m.p.h.
12. 1 h 33 m
13. 1820 miles
14. 10 m
15. £88
16. £105
17. 8 a.m.
18. 1 h 52 m
19. £84
20. £210

UNIT THIRTEEN

A

1. £7.10
2. 219
3. 0.4776
4. 67.5 p
5. 2 mins
6. £13.76
7. 2 h 32 m
8. £150.40
9. 80%
10. 513

B

1. 72.14
2. £180.75
3. £1.63
4. £1.80
5. 24 km/h
6. $62\frac{1}{2}$%
7. 7 h 25 m
8. 4 h 12 m
9. 55
10. £106.09

C

1. £31.14
2. 70 p
3. 60 p
4. 29
5. 732
6. £260
7. £423
8. £92
9. £15
10. £160

D

1. 366
2. 17
3. 11 154 km
4. 220
5. 283.4 francs
6. 80%
7. £3165
8. 9 p
9. £20
10. £128.80

UNIT FOURTEEN

A

1. 38.8666
2. 14.0588
3. 47.535
4. 1.104
5. 97.2196
6. £106.74
7. 20 p
8. 2.468 kg
9. 0.112
10. 7074.754

B

1. 5.438
2. 0.3931
3. 44.44%
4. 0.4762
5. 5.330
6. 5.322
7. 3.996
8. 232.2
9. 2.007
10. 2.883

C

1. 212
2. 45.97 m^2
3. 2.265 m
4. 28.7 m.p.h.
5. 16.66 m^2
6. 4.9 m; £137.00
7. 38.18 m^2
8. 33.18 m^2; 3 tins
9. £14.97
10. 9

D

1. £28.64
2. £466.80
3. £866.04
4. £57.72
5. £6.52
6. £31.68
7. £57.52
8. £85.50
9. £111.30
10. £12.47

E

1.	£4.93	2.	£753.60; £502.40
3.	146.2 km	4.	£45.98
5.	£8.22; £4.70	6.	£161
7.	£19.63	8.	$344.44
9.	£2.58	10.	£282.00

F

1.	3 h 11 m	2.	23.2 m.p.h.
3.	8.32 m.p.h.	4.	15.5 miles
5.	77.5 m.p.h.	6.	1639
7.	29 m.p.h.	8.	6 p
9.	1529	10.	30 m.p.h.

G

1. a) 1215 b) 60.3 m.p.h.
   c) 70 m.p.h.
2. a) 1365 lb
   b) 546 lb : 273 lb : 546 lb
   c) £819
3. a) £589.68 b) 1473 ℓ
   c) £196.56
4. a) £1525.20 b) £227.80
   c) £58.52
5. a) 14.08 gal b) 4.55 ℓ
   c) £1.83
   d) 30.62 miles per gallon

UNIT FIFTEEN

A

1.	73.9%	2.	129%
3.	£25.30	4.	£89.30
5.	44.8%	6.	£92.90
7.	£68.40	8.	16.9%
9.	£20.20	10.	18.7%

B

1.	33.33%	2.	29.44%
3.	£235.60	4.	£24.30
5.	£8.55	6.	£38.25
7.	30.65%	8.	20.00%
9.	£2430	10.	£13.60; 4.69%

C

1.	£9.20	2.	£20.70
3.	£3.05	4.	£91.36
5.	£166.67	6.	£29.74
7.	£17.34	8.	26
9.	£25.42	10.	£11.88

D

1.	£9.49	2.	£18.98
3.	£20.91	4.	£1.83
5.	£27.48	6.	£21.60
7.	£2205	8.	4%
9.	40%	10.	53%

E

1. a) £466.90 b) £104.50
   c) 18%
2. a) £380.16; £381.26; £373.53
   b) £77.56; 25.63% £81.76; £27.30%
   £88.32; 30.97%
   c) £254.89
3. a) £175 500 b) 11.7%
   c) £26 505
4. a) £504 b) £420
   c) 17.5%
5. a) £610; £488; £854
   b) £180.84 c) 464
   d) £3500

UNIT SIXTEEN

A

1.	£120	2.	£49
3.	£287.50	4.	£848.01
5.	£5212	6.	1 year 4 months
7.	6.5%; 6.75%	8.	£2070
9.	Ordinary 8 p	10.	£66.66

B

1.	£233.28	2.	£665.50
3.	£831.67	4.	£2247.88
5.	£6817.33	6.	£56.43
7.	£56.26	8.	£257.82
9.	£209.33	10.	£796.04

C

1. £20 250
2. £22 950
3. £39 568
4. 19.0%; 10.6%
5. £1037

D

1. £24.70
2. £11.20
3. £90.72
4. £102.48
5. $45\frac{1}{2}$ hours
6. £3.24; £4.32; £2.84; £5.68; £3.28; £4.92
7. £110.16
8. £112.18
9. $42\frac{1}{2}$ hours
10. £114.57; £116.67; £166.30

E

1. £33.45
2. £116.42
3. £19.50
4. £857
5. £42.22
6. £6.04
7. £36.76
8. a) £1277.50 b) £5480.93

F

1. £357.30
2. £1042.80
3. £5.55
4. £1603.24
5. £1542.35
6. £2793.60
7. £8222
8. nil
9. £6442.50
10. £44 939.50

G

1. a) £2735.26 b) 8% c) Building Society £255.26
2. a) £14 100 b) £846 c) £8910
3. a) £360 b) £389.91 c) 1.0%
4. a) £1158 b) 15.9% c) £1377
5. a) £4687.80 b) £2724.93 c) £80.90
6. a) £455 b) £21.30; 4.68% c) £609.54

UNIT SEVENTEEN

A

1. £261
2. £222.56
3. £277
4. 114 p
5. £325
6. 2.5 p
7. £13 500
8. a) £31 500 000 b) £26 775 000
9. a) £533.25 b) 6 p
10. a) £29.64 b) 3.5%

B

1. £60.67
2. £136.70
3. Yes; £1.88
4. £345.83
5. £101.07

C

1. £28.87
2. £35.25
3. £74.48
4. £10.81
5. 136
6. 176, 181.6, £63.93, 0, £73.83
7. 1392, £67.23, £73.60, 1323, £60.20, £60.20, £38.57, £172.37
8. £16.15, 728, £32.03, £7.23, £55.41

D

1. a) £55.54 b) £292 c) 736
2. a) 243; £10.45; £23.95; £3.59; £27.54 b) 300
3. a) £100.72 b) £8.45 c) 8.4%
4. a) £40.54 b) £46.62 c) 493
5. a) (i) £149.54; £194.56 (ii) 30.10% b) (i) £101.25 (ii) £127.68 c) £274.36

UNIT EIGHTEEN

A

1. 14.4 cm
2. 15 cm
3. 6 cm
4. 12.2 cm
5. 15.1 cm
6. 7.54 in
7. 5.74 m
8. 3.41 m
9. 4.88 m/s
10. 145 cm

B

1. 125 m^2
2. 415 cm^2
3. 18.8 in
4. 67.0 cm^2
5. 20.0 m
6. 150 cm^2
7. 113 cm^2
8. 886 m^2
9. 643 cm^2
10. 400 m^2; 630.4 m^2; £20.40

C

1. 4.199 m^2
2. 12 ft
3. 7065 cm^3
4. 181 875
5. 0.432 m^2

D

1. 328 cm; 352 cm^3
2. 54 in^2; 27 in^3
3. 220 cm^2; 200 cm^3
4. 5.12 m^2; 0.768 m^2
5. 37.26 m^2; £29.75
6. 240.85 cm; 830
7. 7.64 m; 5.70 m^2; £56.15
8. 160 in^2; 40 in^2

E

1. 24 ℓ
2. 3.0 ℓ
3. mℓ
4. 26 460 ℓ
5. 300
6. 3600 ℓ; 152 cm; 25 min
7. 49.5 ℓ; 7.19 gal; 65.4%; 215.7 miles
8. 47.04 m; 2.82 m

F

1. a) 346.5 cm^2; 283.5 cm^2
   b) 2.772 kg c) 45%
2. a) 3072 ℓ b) 672 gal
   c) 26 min
3. a) 269.5 m^2 b) £443.52
   c) £1.20
4. a) £17.55 b) £135
   c) 77%
5. a) 14.98 m^2 b) £71.88
   c) 30.3%

UNIT NINETEEN

A

1. 70
2. 177
3. 151
4. 14.7%
5. 12.4%
6. 19.8%
7. 56
8. 63, 45, 56, 11
9. 63, 15, 46, 23

B

1. 105°
2. 15°
3. 7 hours
4. 12 hours
5. 12.15 p.m.
6. 29.2%
7. 8.8%
8. £73.92

10. $127\frac{1}{2}°$, $37\frac{1}{2}°$, 60°, $52\frac{1}{2}°$, 60°, $22\frac{1}{2}°$

C

1.  represents 100 stamps
2. 400
3. 8 : 5 : 4 : 7 : 2
4. 320

D

1. £1.20
2. £4.40
3. £2.90
4. £3.00
5. 4.4 kg
6. 6.8 kg
7. 30%
8. £1.50; £1.20; £3.80; £3.00; £3.50
9. £2.60
10. 32.5 miles; 40 p

E

1. a) 167°, 20°, 48°, 125°
   c) £27.69
2. b) £4225 c) £52 048
3. a) £98 b) £96.80
   c) 19%
4. a) 166.5°, 37.5°, 15°, 67.5°, 67.5°, 6°
   c) 345 600 kg
5. a) 4875
   b) week seven; 37 500 c) 18.8%

UNIT TWENTY

1. a) £728 b) 41 387.5
   c) £9.65
2. a) 33.6 m.p.h. b) 4.28 gal
   c) £1.03
3. a) £36.72 b) 3.8 p
   c) 3.02%
4. a) £214.23; £147.82
   b) 3 h 20 min c) 8.8%
5. a) 391 b) £116.80
   c) 1260 in^2
6. a) £541.75 b) £608.16
   c) £116.41
7. a) £1280; £768; £512
   b) £4288 c) £16 723.20
8. a) £500 b) £4786.71
   c) 36.763%
9. a) $\frac{14}{15}$ b) 4.11 kg
   c) butter d) 0.42%

Answers to Tests and Papers

Unit One Test

1. 4854
2. 275
3. 343
4. 730
5. 2600
6. 23, 27
7. 36
8. 1632
9. 41
10. 70

Unit Two Test

1. 23.19
2. 10.39
3. 20.3
4. 24.36
5. 790
6. 0.0854
7. 4.55
8. 16.3
9. 5.25
10. 4

Unit Three Test

1. $\frac{3}{8}$
2. 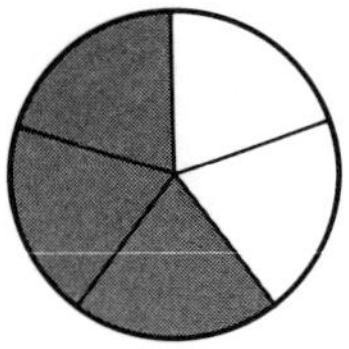
3. $\frac{3}{4}$
4. $\frac{6}{5}$
5. 6
6. $3\frac{8}{15}$
7. $\frac{1}{2}$
8. $4\frac{1}{3}$
9. $\frac{3}{4}$
10. 24

Unit Four Test

1. 2748 p
2. £60.99
3. £102.89
4. 33 p
5. £5.03
6. $1386
7. £260
8. £2.67
9. £43.18
10. 15 p

Unit Five Test

1. 69.2 mm
2. 8.016 kg
3. 7 ft
4. 750 mℓ
5. 12.034 kg
6. 1 ft 1 in
7. 50 000
8. 0.6604 m
9. 20 gal
10. 1.3 m

Unit Six Test

1. 18
2. 3
3. 2.25
4. 9
5. $2\frac{2}{5}$, $2\frac{1}{3}$, 2.33
6. $\frac{11}{2}$, $\sqrt{36}$, 6.1
7. £2.09
8. 1.4
9. 57.6
10. 12 cm

Unit Seven Test

1. $\frac{17}{20}$
2. 175%
3. $6\frac{1}{4}$%
4. £45
5. £920
6. 144
7. £2.20
8. $16\frac{2}{3}$%
9. £88
10. 23%

Unit Eight Test

1. 4
2. 79.17
3. 6
4. £112.50
5. $553
6. £4.24
7. £41.60
8. 149 gal 4 pt
9. 18.2 kg
10. 30%

Unit Nine Test

1.	4 : 13	2.	£1.68
3.	£21 : £35	4.	£630 : £180
5.	£64 : £96 : £192	6.	£30 : £18
7.	45 p	8.	750 g : 500 g : 250 g
9.	84 p	10.	£1250

Unit Ten Test

1.	6.05 cm^2	2.	17.34 cm^2
3.	344 mm^2	4.	180 cm
5.	3 ft	6.	154 cm^2
7.	192 cm^2	8.	6 cm
9.	£10.395	10.	178 yd^2

Unit Eleven Test

1.	21.2 cm	2.	10 m
3.	88 cm	4.	$7\frac{9}{16}$ cm
5.	154 cm^2	6.	100 m
7.	80 cm	8.	0.375 m^3
9.	25 cm	10.	2 cm

Unit Twelve Test

1.	14.40	2.	15 h 45 min
3.	2 h 6 min	4.	8.4 km/h
5.	$6\frac{2}{3}$ miles	6.	$5\frac{5}{9}$ m/s
7.	16 h 40 min	8.	3 h 42 min
9.	£90.27	10.	40 h 40 min

There is no Unit Thirteen Test

Paper One

1.	1748	2.	6000	3.	14
4.	17.89	5.	$112\frac{1}{2}$	6.	$2\frac{5}{16}$
7.	1.6	8.	$1\frac{2}{3}$	9.	30
10.	2495	11.	256	12.	$26\frac{1}{24}$
13.	$1\frac{1}{8}$	14.	$3\frac{1}{3}$	15.	0.375
16.	0.070	17.	$4\frac{1}{3}$	18.	$\frac{1}{2}$
19.	$3\frac{8}{15}$	20.	0.3162	21.	37.13
22.	0.32	23.	$3\frac{3}{20}$	24.	$3\frac{3}{8}$
25.	$1\frac{2}{15}$	26.	1.89	27.	$\frac{3}{88}$
28.	$7\frac{1}{2}$	29.	0.625	30.	1.264

Paper Two

1.	684 p	2.	£2.02
3.	£6.46	4.	£4.20
5.	£1.92	6.	£20
7.	£42	8.	5.255 kg
9.	6 gallons 5 pt	10.	32
11.	15.24 cm	12.	10 in
13.	$1\frac{7}{9}$, 1.88, 1.9	14.	15
15.	3	16.	3
17.	12	18.	$\frac{17}{18}$
19.	1936	20.	$\frac{13}{20}$
21.	0.35	22.	385
23.	$666\frac{2}{3}$	24.	27
25.	8 kg 360 g	26.	£170
27.	94	28.	18
29.	44 p	30.	41 p

Paper Three

1. 1378
2. 7115
3. 25.94
4. 0.9984
5. 0.16
6. $5\frac{19}{20}$
7. $3\frac{1}{12}$
8. 0.0362
9. 960
10. 7200 m
11. 75 : 45
12. 54
13. £7.70
14. £133
15. 3.6
16. $\frac{13}{20}$
17. £200
18. $87\frac{1}{2}\%$
19. 260 cm
20. 21
21. 6 h 24 min
22. £3720
23. 300
24. 20%
25. 43 p
26. 7.45, $7\frac{2}{5}$, $7\frac{1}{3}$
27. 1.1 kg
28. £2500
29. £8.90
30. £36

Paper Four

1. 753
2. 25.09
3. 5092
4. 0.548
5. $7\frac{7}{8}$
6. 4
7. 0.006
8. 8
9. £90.72
10. £2.40
11. 2765
12. 35 m 42 cm
13. 114
14. 1.37, $1\frac{3}{7}$, 1.73
15. £37.50
16. 90%
17. $\frac{3}{8}$
18. $127.5
19. 1.043 kg
20. 70 miles
21. 45 p : 30 p
22. 49 cm^2
23. 4 h 50 min
24. 2600 francs
25. 50
26. 21, 26
27. £1.98
28. 310
29. 154 cm^2
30. £93

Paper Five

1. 1420
2. 17.68
3. 143
4. 6590
5. $2\frac{7}{12}$
6. 30
7. 1.7
8. 875 g
9. £156.75
10. £21.60
11. 10
12. 3 : 4
13. 120
14. 0.26, $\frac{1}{4}$, 23%
15. £204
16. 40 m
17. 20
18. 11.25 cm^2
19. 1 ft $1\frac{1}{4}$ in
20. 6 ℓ
21. £75 : £100
22. 22 m.p.h.
23. 5 h 6 m
24. $1274
25. 1 cm
26. 16, 19
27. £1122
28. 22 m
29. 616 cm^2
30. 19 Nov

Paper Six

1. 214
2. 868
3. 9
4. $2\frac{7}{8}$
5. 2
6. £15.19
7. £1.45
8. 80 cm
9. 21 kg 200 g
10. £2.45
11. 17
12. 12
13. 151.2
14. £100 : £75
15. 250 g
16. 5.02
17. £28.35
18. 0.74, $\frac{3}{4}$, 76%
19. 28.8 cm
20. £12
21. 3 h 31 min
22. £40.25
23. 200
24. 88 cm
25. £95.88
26. £1095
27. 1080 km
28. £50
29. 123 kmph
30. 4.50 p.m.

Unit Fourteen Test

1. a) 25 kg : $12\frac{1}{2}$ kg : $12\frac{1}{2}$ kg b) 69 p
   c) 17.2%
2. a) 38 994 b) £4.92
   c) £2.20
3. a) 21.15 b) £115.92
   c) 87%
4. a) £15 892
   b) £17 308.56
   c) £118.60; £129.17
5. a) £304.60 b) 38.5%
   c) £38.20 cheaper, 12.5%

Unit Fifteen Test

1. a) £9300 b) £1860 c) 62.9%
2. a) £157.35 b) £226 c) £204
3. a) £60 b) £210
4. a) £6.50 b) 26%

Unit Sixteen Test

1. a) £4445 b) £4158.35
   c) A £3.15
2. a) £243.75; £262.50; £281.25
   b) 15.4%
   c) £5664.40
3. £319

Unit Seventeen Test

1. a) £676.00 b) £405.60; £270.40
2. £24.67
3. a) £1489.43 b) 18.6 p
   c) 33%
4. a) £38 500 b) 5 p

Unit Eighteen Test

1. a) 6600 cm^2 b) 4668 cm^2
   c) 19 152 cm^3 d) £5.36
2. a) 2772 ℓ b) 606.375 gal
   c) 18 minutes
3. 766
4. a) 5.66 b) 96 cm^3
   c) 180 cm^3
5. a) 113 cm^3 b) 102 mℓ
   c) 19

Unit Nineteen Test

3. 50, 60, 75, 100, 90, 130
4. a) 6 b) 23 c) 56.5%

Paper Seven

1. a) £448.88 b) £440.89
   c) Bank £7.99
2. a) £9.18 b) £5380
3. a) £2.45 b) $43\frac{1}{2}$
4. a) £892.80 b) £127.28
5. a) 346.25 b) £554.63
6. 30, $52\frac{1}{2}°$, $22\frac{1}{2}°$, 45°, 210°
7. a) £79.56 b) 289

Paper Eight

1. a) £89.50 b) £82.50
   c) £8.50, 10.3%
2. a) £2668.68; 11.14%
   b) £3166.20; 11.66%
   c) £3905.10; 12.7%
   d) £1041.75

3. a) £690.86 b) £3634.86
   c) £22.65

4. a) 1244 m^2 b) £195
   c) 39 d) £882.17

5. a) 23 b) 7.17 seconds
   c) 34.8%

6. a) 15.14 b) 43.4 m.p.h.
   c) £161.60

Paper Nine

1. a) £112.50; £126.56
   b) £1580
   c) £1732.50; £155.07

2. a) 77 cm^2 b) 161 cm^2
   c) 64.38 cm

3. a) £61 800 b) £116.85
   c) £1284.49

4. a) 2472 b) £104.34
   c) £31.34

5. a) £447 m, 270°; £99 m, 60°; £50 m, 30°
   b) £908 m, £125 m

6. a) 13 m.p.h. b) 11.07 a.m.

7. a) £15.80 b) £6.60
   c) £3 A

Index